MISCELLANEOUS PROBLEMS IN MARITIME NAVIGATION, TRANSPORT AND SHIPPING

Miscellaneous Problems in Maritime Navigation, Transport and Shipping

Marine Navigation and Safety of Sea Transportation

Editors

Adam Weintrit & Tomasz Neumann
Gdynia Maritime University, Gdynia, Poland

CRC Press
Taylor & Francis Group
Boca Raton London New York

CRC Press is an imprint of the
Taylor & Francis Group, an **informa** business

A BALKEMA BOOK

CRC Press/Balkema
P.O. Box 447, 2300 AK Leiden, The Netherlands
e-mail: Pub.NL@taylorandfrancis.com
www.crcpress.com – www.taylorandfrancis.com

First issued in hardback 2017

ISBN-13: 978-0-4156-9118-5 (pbk)
ISBN-13: 978-1-1384-3578-0 (hbk)

Visit the Taylor & Francis Web site at
http://www.taylorandfrancis.com

and the CRC Press Web site
http://www.crcpress.com

List of reviewers

Prof. Roland **Akselsson**, Lund University, Sweden,
Prof. Yasuo **Arai**, President of Japan Institute of Navigation, Japan,
Prof. Michael **Barnett**, Southampton Solent University, United Kingdom,
Prof. Tor Einar **Berg**, Norwegian Marine Technology Research Institute, Trondheim, Norway,
Prof. Alfred **Brandowski**, Gdańsk University of Technology, Gdynia Maritime University, Poland,
Prof. Zbigniew **Burciu**, Master Mariner, Gdynia Maritime University, Poland,
Prof. Shyy Woei **Chang**, National Kaohsiung Marine University, Taiwan,
Prof. Adam **Charchalis**, Gdynia Maritime University, Poland,
Prof. Krzysztof **Chwesiuk**, Maritime University of Szczecin, Poland,
Prof. Krzysztof **Czaplewski**, Polish Naval Academy, Gdynia, Poland,
Prof. Andrzej **Felski**, President of Polish Navigation Forum, Polish Naval Academy, Gdynia, Poland,
Prof. Wlodzimierz **Filipowicz**, Master Mariner, Gdynia Maritime University, Poland,
Prof. Masao **Furusho**, Master Mariner, Kobe University, Japan,
Prof. Avtandil **Gegenava**, Batumi Maritime Academy, Georgia,
Prof. Witold **Gierusz**, Gdynia Maritime University, Poland,
Prof. Stanislaw **Gorski**, Master Mariner, Gdynia Maritime University, Poland,
Prof. Lucjan **Gucma**, Maritime University of Szczecin, Poland,
Prof. Michal **Holec**, Gdynia Maritime University, Poland,
Prof. Qinyou **Hu**, Shanghai Maritime University, China,
Prof. Marek **Idzior**, Poznan University of Technology, Poland,
Prof. Mirosław **Jurdzinski**, Master Mariner, FNI, Gdynia Maritime University, Poland,
Prof. Lech **Kobylinski**, Polish Academy of Sciences, Gdansk University of Technology, Poland,
Prof. Krzysztof **Kolowrocki**, Gdynia Maritime University, Poland,
Prof. Serdjo **Kos**, FRIN, University of Rijeka, Croatia,
Prof. Eugeniusz **Kozaczka**, Polish Acoustical Society, Gdansk University of Technology, Poland,
Prof. Andrzej **Krolikowski**, Master Mariner, Maritime Office in Gdynia, Poland,
Prof. Pentti **Kujala**, Helsinki University of Technology, Helsinki, Finland,
Prof. Jan **Kulczyk**, Wroclaw University of Technology, Poland,
Prof. Bogumil **Laczynski**, Master Mariner, Gdynia Maritime University, Poland,
Prof. Andrzej **Lewinski**, Radom University of Technology, Poland,
Prof. Mirosław **Luft**, President of Radom University of Technology, Poland,
Prof. Zbigniew **Lukasik**, Radom University of Technology, Poland,
Prof. Artur **Makar**, Polish Naval Academy, Gdynia, Poland,
Prof. Aleksey **Marchenko**, University Centre in Svalbard, Norway,
Prof. Torgeir **Moan**, Norwegian University of Science and Technology, Trondheim, Norway,
Prof. Wacław **Morgas**, Polish Naval Academy, Gdynia, Poland,
Prof. Nikitas **Nikitakos**, University of the Aegean, Greece,
Prof. Wiesław **Ostachowicz**, Gdynia Maritime University, Poland,
Mr. David **Patraiko**, MBA, FNI, The Nautical Institute, UK,
Prof. Vytautas **Paulauskas**, Master Marine, Maritime Institute College, Klaipeda University, Lithuania,
Prof. Francisco **Piniella**, University of Cadiz, Spain,
Prof. Marcin **Plinski**, University of Gdansk, Poland,
Prof. Chaojian **Shi**, Shanghai Maritime University, China,
Prof. Leszek **Smolarek**, Gdynia Maritime University, Poland,
Prof. Jac **Spaans**, Netherlands Institute of Navigation, The Netherlands,
Prof. Cezary **Specht**, Polish Naval Academy, Gdynia, Poland,
Cmdr. Bengt **Stahl**, Nordic Institute of Navigation, Sweden,
Prof. Anna **Styszynska**, Gdynia Maritime University, Poland,
Prof. Janusz **Szpytko**, AGH University of Science and Technology, Kraków, Poland,
Prof. Elżbieta **Szychta**, Radom University of Technology, Poland,
Prof. Mykola **Tsymbal**, Odessa National Maritime Academy, Ukraine,
Prof. Waldemar **Uchacz**, Maritime University of Szczecin, Poland,
Prof. Dang Van **Uy**, President of Vietnam Maritime University, Haiphong, Vietnam,
Prof. Peter **Voersmann**, President of German Institute of Navigation DGON, Deutsche Gesellschaft für Ortung und Navigation, Germany,
Prof. Vladimir **Volkogon**, Rector of Baltic Fishing Fleet State Academy, Kaliningrad, Russia,
Prof. Adam **Weintrit**, Master Mariner, FRIN, FNI, Gdynia Maritime University, Poland,

Contents

Miscellaneous Problems in Maritime Navigation, Transport & Shipping

Introduction

A. Weintrit & T. Neumann
Gdynia Maritime University, Gdynia, Poland

PREFACE

The contents of the book are partitioned into seven parts: weather routing and meteorological aspects (covering the chapters 1 through 6), ice navigation (covering the chapters 7 through 9), ship construction (covering the chapters 10 through 15), ship propulsion and fuel efficiency (covering the chapters 16 through 21), safe shipping and environment in the Baltic Sea Region (covering the chapters 22 through 26), oil spill response (covering the chapter 27 through 29), large cetaceans (covering the chapters 30 through 32).

The first part deals with weather routing and meteorological aspects. The contents of the first part are partitioned into six chapters: Elements of tropical cyclones avoidance procedure, Baltic navigation in ice in twenty first century, Storm-surges indicator for the Polish Baltic Coast, Polish seaports – unfavorable weather conditions for port operation (applying methods of complex climatology for data formation to be used by seafaring), Analysis of hydrometeorological characteristics in Port of Kulevi zone, and Hydrometeorological characteristics of the Montenegrin coast.

The second part deals with ice navigation. The contents of the second part are partitioned into three chapters: Ship's navigational safety in the Arctic unsurveyed regions, Methods of iceberg towing, Ice management – from conception to realization.

The third part deals with ship construction. The contents of the third part are partitioned into six chapters: Investigations of marine safety improvements by structural health monitoring systems, Ultrasonic sampling phased array testing as a replacement for X-ray testing of weld joints in ship construction, Conditions of carrying out and verification of diagnostic evaluation in a vessel, Determination of ship's angle of dynamic heel based on model tests, Propulsive and stopping performance analysis of cellular container carriers, and Coales-cence filtration with an unwoven fabric barrier in oil bilge water separation on board ships.

The fourth part deals with ship propulsion and fuel efficiency. The contents of the fourth part are partitioned into six chapters: Optimization of hybrid propulsion systems, Integrating modular hydrogen fuel cell drives for ship propulsion: prospectus and challenges, Modeling of power management system on ship by using Petri Nets, Logical network of data transmission impulses in journal-bearing design, Optimum operation of coastal merchant ships with consideration of arrival delay risk and fuel efficiency, and Digital multichannel electro-hydraulic hxecution improves the ship's steering operation and the safety at sea (security of the navigation act).

The fifth part deals with safe shipping and environment in the Baltic Sea Region. The contents of the fifth part are partitioned into five chapters: Towards the model of traffic flow on the Southern Baltic based on statistical data, Incidents analysis on the basis of traffic monitoring data in Pomeranian Bay, Model of time differences between schedule and actual time of departure of sea ferries in the Świnoujście Harbour, Simplified risk analysis of tanker collisions in the Gulf of Finland, and Estimating the number of tanker collisions in the Gulf of Finland in 2015.

The sixth part deals with oil spill response. The contents of the fifth part are partitioned into three chapters: The method of optimal allocation of oil spill response in the Region of Baltic Sea, Modeling of accidental bunker oil spills as a result of ship's bunker tanks rupture - a case study, and The profile of Polish oil spill fighting system.

The seventh part deals with large cetaceans. The contents of the seventh part are partitioned into three chapters: Towards safer navigation of hydrofoils: avoiding sudden collisions with cetaceans, Estimation on audibility of large cetaceans for improvement of the under water speaker, and Feasibility study on infrared detecting of large cetaceans to avoid sudden collisions.

Weather Routing and Meteorological Aspects

1. Elements of Tropical Cyclones Avoidance Procedure

B. Wiśniewski & P. Kaczmarek

Maritime University of Szczecin, Poland

ABSTRACT: The updated version of the Cyclone II program was used for analyzing hundreds of cases where ships were facing dozens of developed cyclones. The program generates directions for navigators that are recommended for consideration before making decisions on passing around or avoiding tropical cyclones. Three specific situations were defined where a vessel may enter the area affected by a tropical cyclone, and its commander must consider three recommendations for safe passing of the cyclone:
– vessel – cyclone encounter, where if on opposite course, the most effective is course alteration;
– when the ship overtakes the cyclone, speed reduction is the most effective action;
– when the vessel and the cyclone are on crossing routes ($30 \div 90°$), a slight decrease in speed or a slight course alteration or both actions can be effective.

1 INTRODUCTION

In areas where tropical cyclones occur navigators must apply the procedure for obtaining information and the identification of the danger on the expected track of the vessel, the accuracy in determining the sectors safe and unsafe, taking into account velocity vectors of the ship (V_s) and the cyclone (V_c). As a result, we decide how to most effectively avoid the area threatened by a tropical cyclone [1-5].

The means for using the procedure include a computer program developed by these authors, described in previous publications [6]. Among others, the software takes into account the procedure for programming the route of the vessel at the time of receiving the message about the cyclone for the following hours and days, with forecasts of expected cyclone positions, and predicted by calculations future positions of the ship.

2 PROCEDURES AND METHODOLOGY

One may find that being together with the cyclone vessel is possible in three specific situations:
– vessel and cyclone move in opposite directions (ahead or nearly ahead of the ship);
– vessel catches up with the cyclone on a similar course;
– projected route of the ship crosses the projected path of the cyclone (eye of the cyclone track) at an angle of ($30° \div 90°$) and the moment when the

ship will enter the waters on tropical cyclone threat.

For these three variants examples were tested, based on real data on cyclones in the years 2008 - 2010 and for the actual and random positions of selected ships owned by the Polish Steamship Company (PSC). Typically, for one cyclone four vessels in positions spaced symmetrically around the cyclone were chosen, that is, from the point of dangerous winds ≥ 34 kn. The moment of decision-making was the situation when calculations showed the entry of the vessel in a dangerous sector of the cyclone in less than 48 hours (TCPA)

The test results will be illustrated by a situation, where the cyclone Bill was being avoided from 20 to 23 August 2009. The cyclone moved along a parabolic trajectory across the North Atlantic. Part of the first message of 20 Aug 2009 is shown in Table 1.

Table 1. Fragment of a message printout: cyclone BILL on 20/08//2009

```
ZCZC MIATCMAT3 ALL
TTAA00 KNHC DDHHMM CCA
HURRICANE BILL FORECAST/ADVISORY NUMBER
21...CORRECTED
NWS TPC/NATIONAL HURRICANE CENTER MIAMI FL
AL032009
1500 UTC THU AUG 20 2009

REPEAT...CENTER LOCATED NEAR 22.6N 61.7W AT
20/1500Z
AT 20/1200Z CENTER WAS LOCATED NEAR 22.1N
61.0W
```

FORECAST VALID 21/0000Z 24.2N 63.8W
MAX WIND 110 KT...GUSTS 135 KT.
64 KT... 90NE 45SE 30SW 75NW.
50 KT...120NE 90SE 60SW 100NW.
34 KT...225NE 200SE 100SW 200NW.

FORECAST VALID 21/1200Z 26.6N 66.0W
MAX WIND 115 KT...GUSTS 140 KT.
64 KT... 75NE 45SE 30SW 45NW.
50 KT...120NE 100SE 70SW 100NW.
34 KT...225NE 200SE 120SW 180NW.

FORECAST VALID 22/0000Z 29.5N 67.5W
MAX WIND 115 KT...GUSTS 140 KT.
64 KT... 75NE 45SE 30SW 45NW.
50 KT...120NE 100SE 70SW 100NW.
34 KT...225NE 200SE 120SW 180NW.

FORECAST VALID 22/1200Z 32.5N 69.0W
MAX WIND 110 KT...GUSTS 135 KT.
50 KT...120NE 100SE 70SW 100NW.
34 KT...225NE 200SE 120SW 180NW.

FORECAST VALID 23/1200Z 40.5N 66.5W
MAX WIND 100 KT...GUSTS 120 KT.
50 KT...120NE 120SE 70SW 100NW.
34 KT...225NE 225SE 120SW 180NW.

Four ship positions for 12.00 (20.08.2009)
- a ship at the port road of New York travelling to Brazil,
- vessel B in the position $\varphi = 15.07°N$, $\lambda = 059.9°W$ on the way to N. York,
- ship in the Mona Passage $\varphi = 18.04°N$, $\lambda = 074.98°W$ on the way to Europe,
- ship in the actual position D $\varphi = 40°N$, $\lambda = 040°W$ on the way to N. York,
- and the position of the cyclone $\varphi c = 22.1°N$, $\lambda = 061°W$ at the same time and day.

In connection with weather forecasts for the next 12, 24, 36, 48, and 72 hours, cyclone positions were taken from messages and then future ship's positions were calculated.

3 RESULTS

For ship (A) departing from New York after receiving a message about the cyclone, data were entered to the computing program "Cyclone II ":
- their position, intended course (course over ground = 15°) to Brazil, and the estimated speed of 13,
- data on the location of the cyclone, its course and speed (vertical panel) - Fig.1.,
- data on forecasts of the cyclone 12, 24, 36, 48, 72 hours (horizontal panel) - Fig.1.

The calculation results indicate that the ship proceeds to a dangerous course sector 149° - 197°, 1254.7 Nm from the eye of the cyclone, and after 43.7 hours (TCPA) it will reach the closest point approach 234Nm (CPA). By obtaining the cyclone position from 72 hour forecast illustrating the predicted cyclone movement path, the captain chooses a new ship's course (COG = 180°) and decides to check what the vessel and cyclone positions will be in 12, 24, 36, 48 and 72 hours. Figure 2 illustrates the results of calculations and relative ship – cyclone positions after 24 hours. The ship passes the cyclone after 48 hours and alters the new COG = 130° to return on route to Brazil, being safe on this course at a distance of 328.4 Nm from cyclone eye (Fig. 3). Finally, the ship extends its distance covered by 210Nm, maintaining a speed of 13.2 kn, i.e. and will prolong the voyage by 16 hours to pass by the cyclone.

For ship (B) in position: 51.07°N, 059.9°W on the way to New York (COG = 230°, Vs = 23kn) it is estimated that it is on a dangerous course (308 ÷ 343°) and after 37.8 hours will be at the closest point approach, i.e. 129.2 Nm from the cyclone influence, affected by winds ≥ 34 knots (Fig. 4). The ship has to reduce its speed below 19.1 kn.

To avoid entry into the cyclone-affected area, the vessel loses 72Nm (3 x 24), which means proceeding at a speed of 19.1kn. This prolongs the expected time of voyage by about 3 hrs 40 min.

Ship (C), sailing in the Mona Passage on its way to Europe, performs testing to pass by the cyclone Bill for the same times on 20.08.2009. The information obtained is that it is on the boundary of dangerous sector (331° to 55°) remaining on course COG = 055° and sailing at a speed of 13 knots (results in the vertical panel - Figure 5). By introducing cyclone data forecast to the program Cyclone II for up to 72 hours, the ship commander finds out that continuing the trip at a speed Vs = 13.0kn, after 12 hours (Fig. 6) its COG = 055°, while the ship will get into a safe sector reaching the closest point of approach 333.4 Nm from the cyclone eye. For this situation no changes of speed and course relative to the cyclone Bill are needed, therefore neither distance not voyage time will be extended.

On 20.08.2009. at 12.00UTC ship (D) is very far from the cyclone, in position $\varphi = 40°N$, $\lambda = 039.15°W$ and on course COG = 270° on the way to New York. From entered forecast data and expected vessel positions (Vs = 13.0kn), only simulation (testing) for 48 hours shows that ship's course will approach the dangerous sector (272°-358°), then sailing at a distance of 907.16 Nm from the cyclone eye (Fig. 7). A simulated situation for passing the cyclone Bill is shown in Figure 7, 8 and 9 for August 22nd 2009. The ship, sailing one day on course COG = 230° and then returning to the course COG = 288° leading to New York, extended the original rhumbline route by about 104 Nm only, corresponding to a prolonged travel time of 8 hours. The ship made a successful maneuver, passing the cyclone at a distance of 338Nm from the outer cyclone dangerous area, where wave heights were 4.0m ≥ m. Performing simulations of the cyclone and ship position projections for the next 72 hours, we get a positive result confirming cyclone avoidance.

4 CONCLUSIONS

1 For the vessel and the cyclone collision situation on opposite courses (ship A) the course alteration is the most effective. The reduction of speed only did not give a positive result for a speed of cyclone movement Vc ≥ 10kn.
2 For the situation when the ship overtakes the cyclone, the most effective is ship's speed reduction relative to cyclone movement, because the track will not be prolonged, and is fuel consumption is likely to be lower (ship B).
3 A special case is where a vessel expects that its planned route will cross the cyclone path, but without the entry of the vessel into the cyclone affected area, and it does not alter change speed or course (vessel C).

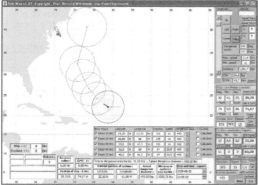

Figure 1 Graphic illustration of the cyclone path, based on weather report and calculations for ship A, heading for Brazil (20-08-2009, 12.00UTC).

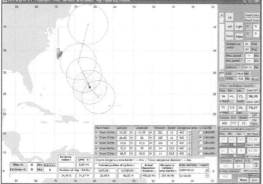

Figure 2 Graphic picture of cyclone and ship A positions, after 24 hour travelling time, and calculation results.

4 With the projected route of the vessel and the cyclone path are likely to cross each other (30 ÷

90°), and when the probability of vessel entry into the area of cyclone influence, slight ship's speed decrease or small course alteration (ΔCOG ≤ 40°) can be effective, as illustrated by the case of ship D, or both changes mentioned above may be made.

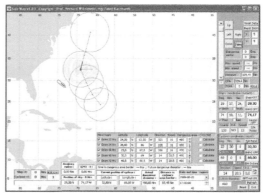

Figure 3 Graphic picture of ship A and cyclone positions after 48 hours in relation to course alteration to 130 degrees.

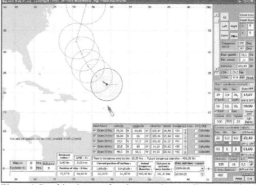

Figure 4 Graphic picture of cyclone track according to received data and calculations for ship B, heading for New York (20-08-2009, 1200 UTC).

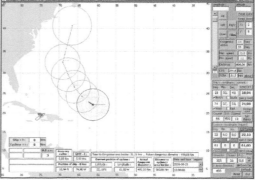

Figure 5 Graphic picture of cyclone track according to received data and calculation results for ship C, Mona Passage, ship proceeding to Europe (20-08-2009, 1200 UTC).

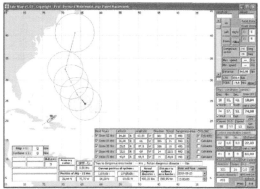

Figure 6 Graphic picture of cyclone and ship C positions during the voyage and calculation results.

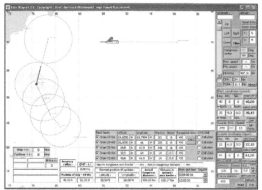

Figure 7 Graphic results of cyclone track according to message and calculation results for ship D after 48 travel hours from 20-08-2009, at. 1200 UTC.

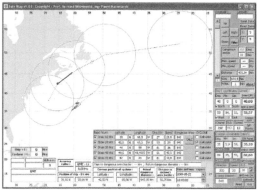

Figure 8 Graphic picture of cyclone and ship D positions according to the message on 22-08-2009, after altering course.

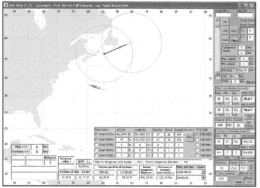

Figure 9 Graphic picture of cyclone and ship D positions after 12 hours after altering course to 288 degrees, voyage to to New York.

REFERENCES

[1] Chomski J., Wiśniewski B., Medyna P. Analysis of ship routes avoiding tropical cyclones. Sympozjum Nawigacyjne. Wyd. AMW Gdynia 2008.
[2] Medyna P., Wiśniewski B., Chomski J. Methods of avoiding tropical cyclone of hurricane Fabian. Scientific Journals Maritime University of Szczecin 2010, 20(92) p.p.92-97.
[3] Wiśniewski B. Radio fax Charts in sea navigation (in Polish).
[4] Wiśniewski B., Potoczek W., Chołaściński A. Określenie sektora kursów niebezpiecznych przy omijaniu cyklonu tropikalnego z wykorzystaniem programu komputerowego. Budownictwo Okrętowe i Gospodarka Morska, nr 11-12/1990.
[5] Wiśniewski B., Chomski J., Drozd A., Medyna P. Omijanie cyklonu tropikalnego w żegludze oceanicznej. Inżynieria Morska i Geotechnika, nr 5/2001, s.296-300.
[6] Wiśniewski B., Kaczmarek P. Avoidance of tropical cyclones using the Cyclon II program. Scientific Journals Maritime University of Szczecin 2010, 22(94) pp.71-77.

2. Baltic Navigation in Ice in the Twenty First Century

M. Sztobryn

Institute of Meteorology and Water Management - National Research Institute, Maritime Branch, Gdynia, Poland

ABSTRACT: The Baltic Sea, even though not large in the global scale, is an important shipping lane. In winter, especially in the region of the Gulf of Bothnia, navigation is seriously obstructed by ice. The aim of this work was investigation of changes in the intensity of the obstruction by ice, caused by climate change in coming 90 years of the twenty first century. It was one of the first attempts of technical application of the global climate scenarios effects. It should be stressed, that the presented work results (as application of the climate scenarios), couldn't be treated as forecast, as it is only the changes tendency assessment.

The climate changes were examined as the changes in air temperature (adaptation of global emission models to regional scale) and atmospheric pressure gradient (model ECHAM5), according to three global scenarios B, A2 and A1B. The number of cases, in which Swedish and Finnish icebreakers assisted the ships, was assumed as the indicator of navigation obstruction by ice "K".

The severity of sea ice conditions was presented by the indicator "S", calculated as the mean value of regional indices. The "S" is the function of the number of days with sea ice, observed at the stations in the particular regions and probability of the sea ice appearances.

Relations between sea ice severity index "S" and regional climate parameters (monthly and annual air temperature and atmospheric pressure gradient) were calculated for the calibration period of 1956-2004. Three models were build: model 1a. "thermal" (the dependence on the mean monthly temperature of July and December); model 1b. thermal "B" (the function of average annual air temperature and of the mean monthly December temperature) and model 2 "thermal zonal" (the dependence on the mean temperature of July and December and zonal component of air pressure gradient). The level of approximation was similar for the analyzed models (over 0,6). Calculation of the future (in XXI century) changes of indicator "K" was done according to three scenarios B, A1B and A2. The number of icebreakers' assistance events should be lower than the one in the twentieth century. The lowest intensity of this decrease is estimated by model 2 and scenario A1Br1, the highest one – for model M1b and scenario B r1.

Otherwise, the minimum value, calculated for the scenarios, is higher than in a period of 1956-2004. It means, that probably, the period with obstruction for navigation in ice could be longer, but not as severe as in the period of 1956-2004. The obstruction intensity could increase during the 21 century according to scenario B1r1, the same for empirical model 2. The similar tendency has been shown by scenario A1Br1 and by model 1a. Other models and scenarios estimated the decreasing trend up to 2100

1 INTRODUCTION

The Baltic Sea, even though not large in the global scale, is an important shipping lane. In winter, especially in the region of the Gulf of Bothnia, navigation is seriously obstructed by ice. Operation of icebreakers to support maritime trade is usually included in the infrastructure offered by the port states. Only in Russia private icebreaker services are operating, as, for example, the Gazprom icebreakers.

On the area of the Gulf of Bothnia – the severe region of the Baltic Sea, in general, Swedish and Finnish icebreakers are in charge.

The aim of this work was investigation of changes in the intensity of navigation obstruction by ice, occurring in effect of climate changes in coming 90 years of the twenty first century.

The observed in XX century changes of climate, especially a rise of the mean air temperature induced more intensive interest in climate modeling and forecasting future changes thereof within a long lasting time period, e.g. by the end of XXI century. From among many scientific research institutions engaged in this subject, the most spectacular achievements (Nobel prize) were gained by IPCC (Intergovernmental Panel on Climate Change). The works of IPCC have been published in a form of reports. According to the reports, it seems to be the most likely, that a cause of the observed climate changes in XX century was the anthropogenic growth of greenhouse gas concentration. The Special Report on Emissions Scenarios (SRES) has presented probable scenarios of greenhouse gas emissions in XXI century. Thus, the predicted gas emission and a rise of air temperature, calculated on the grounds of the scenarios, were used as the input data to the global climate models – the global circulation models - (among the others: ECHAM model for Europe and the North Atlantic).

In the research there have been used scenarios B1, A2 and A1B of greenhouse gases emission, worked out by IPCC for XXI century. The differences between the scenarios result from varying assumptions of the world evolution in XXI century, due to globalization, economic development, predicted changes of population and also abiding by the sustainable development principles.

Thus, in B1 scenario, an very integrated global development with simultaneous implementation of the sustainable development was assumed. It would cause a rapid economic growth focused on services and informatics sciences. By 2050 a growth of population and next its drop are expected.

According to A2 scenario the much less integrated development was assumed (the independently developing regions /nations will be more significant). It will result in fast and continuous growth of people, free economic development focused on regional benefits, less and more diversified development, differentiated abiding by sustainable development (to the total denial) and minor changes in technology.

Furthermore, in A1B scenario there was assumed a rapid economic growth with the balanced exploitation of all accessible power supply sources (fossil and non fossil ones) with introducing the new, more productive technologies as well. It has to be emphasized that the European Union accepted A1B scenario as a basis for shaping the energy policy.

Adaptation of the emission scenarios (A2, B1 and A1B) results, also the results of the global climate model ECHAM to the Baltic and Poland conditions was carried out within the framework of KLIMAT Project (www.klimat.imgw.pl) by two IMGW-PIB Departments : Zakład Modelowania Klimatycznego i Prognoz Sezonowych oraz Centrum Monitoringu Klimatu Polski

The basis included re-analyzing and downscaling performed for the selected reference period (1955-2004) between the global driving factor (the same climate elements and the same spatial grid as in SRES and ECHAM models) and the regional one. The obtained relations between the global and the regional driving factor were used to create the regional scenarios for the Baltic Sea and Poland in XXI century; the global driving factor was represented by SRES and ECHAM models' results. Adaptation of the global models to the Baltic and Poland conditions and working out the regional driving factor field are results of the works carried out by the IMGW PIB and were described, among the others, in the work by Jakusik (Jakusik at al. 2010a, Jakusik at al. 2010b), also in the reports of Klimat project presented on the web page klimat.imgw.pl.

Two elements of the regional driving factor field were adopted for purposes of the research: monthly air temperature and the component indicators (meridian and zonal) of the atmospheric circulation (monthly, annual). In this case as well, for the calibration period there were constructed statistic and empiric models, displaying relationship between the selected parameters (Baltic sea-ice severity index and implicit a number of icebreakers assistance events) and the predictors, which are characteristic for the regional driving factor field i.e. the monthly and annual air temperature and the component indicators (meridian and zonal) of the atmospheric circulation (monthly, annual). The obtained statistic and empiric relations and results of the scenarios for the regional driving factor have been used for assessment of changes in the XXI century.

The algorithmic description of methodical actions performed during work realization is shown in Fig. 1.

2 DATA

In the calculations 3 types of data were applied: results of the regional driving factor field, sea-ice severity index and a number of the icebreakers assistance events occurred in the Baltic Sea.

Regional driving factors

The following elements were subject to analyzing: monthly air temperatures (global greenhouse gases emission scenarios downscaling – SRES models) and the atmospheric circulation (result of ECHAM model downscaling) - component indicators (meridian and zonal) of the atmospheric circulation (monthly, annual).

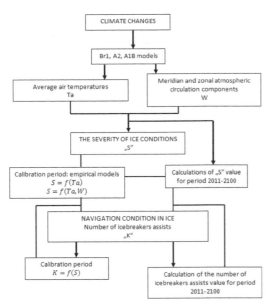

Figure 1. The algorithmic description of methodical actions performed during work realization.

In the following part of the work, dedicated to modeling of sea-ice conditions for various scenarios, an applied scenario summary was presented together with the model name. Thus for example, a name of model M1 A2 designates M1 model, where the original source of driving factors had been A2 scenario.

Sea Ice Severity index

Sea ice severity index „S" originally was constructed for Polish Coastal zone (Southern Baltic) in 2006 (Sztobryn M. 2006) and next, in 2009, was applied to determine conditions of the whole Baltic sea ice condition (Sztobryn and others 2008) . The index is based on the probability of ice occurrence and the number of days with sea ice observed in the analyzed period in the selected basins. The ice severity index is given by the relation

$$S = 0.05 \times \frac{1}{i} \sum_j \left(\frac{N}{p} \right)_j$$

where

S - severity index,

N - number of days with ice observed during the winter season at the particular station,

p - probability of ice occurrence at the particular station, calculated for the analyzed period and

i - number of stations taken into account.

To calculate the Baltic Sea Ice Severity there were applied the data referring to numbers of days with ice and its occurrence probability taken from 34 stations from 1955 to 2004 (Western Baltic Sea, with: Unterwarnow, Warnemünde, Kiel LH; Southern Baltic Sea, with: Zalew Szczeciński, Świnoujście, Kołobrzeg, Gdańsk, Zalew Wiślany; Gulf of Finland, with: Hanko, Russarö, Helsinki. Harmaja, Helsinki LH, Loviisa, Oerengrund, Hogland; Sea of Aland and the Archipelago, with: Maarianhamina, Koppaklintar, Lågskär, Turku, Bogskär (Kihti), Utö Sea of Bothnia, with: Rauma, Kylmäpihlaja, Raumanmatala , Norra Kvarken, with: Haasa, Ensten, Norrskär, Bay of Bothnia, with: Ajos, Mutkanmatala, Kemi One, Ykspihlaja, Repskär, Tankar).Variability of sea ice severity index and possibility of application thereof in prediction and modeling of influence of sea ice conditions on navigation, including also icebreakers activity were presented in 2009 at TRANSNAV conference (Sztobryn and others 2009).

Icebreakers activity – navigation in sea ice

In this work it was assumed that the indicator of difficulties "K" in the navigation related to the occurrence of ice phenomena is the number of assistance events wIth Swedish and Finnish icebreakers, during the ice season. The data including a number of icebreakers assistance events in specific seasons come from the annual SMHI Works: Report of sea ice condition and icebreakers' activity.

3 RELATION BETWEEN THE ICEBREAKERS ASSISTANCE EVENTS NUMBER (K) AND SEA ICE SEVERITY INDEX (S)

Obstructions in navigation on the Baltic Sea while ice cover occurrence were represented with the indicator of difficulties in navigation, which was equal to a number of Swedish and Finnish icebreakers assistance events within a specific ice season. There has been analyzed and next constructed the mathematic model.

The relationship between ice condition (represented by "S" index) and navigation condition in ice (by number of icebreakers' assists) was formulated by the exponential regression method.

The number of icebreakers assists, recorded in 1956-2004 period, was given by the formula 1 :

$$K = a \times \exp(b \times S) \tag{1}$$

where:

K – calculated number of assistance events,

S – indicator of the Baltic sea ice severity (mean),

a,b, - numerical coefficients

Comparison of the course of the calculated number of icebreakers assistance events and the real number thereof in 1956-2004 is shown on Fig. (2). The black, thin line stands for the real number of icebreakers assistance events, whereas the grey thick one represents the number of events calculated using the formula.

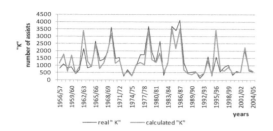

Figure 2. Comparison of the calculated and the real numbers of icebreakers assistance events in 1956-2004 period.

Differences between the "K" indicator real value (a number of assistance events) and the one calculated using formula 1 ,- are specially visible for calculations performed for 1995/96 season. However it is characteristic for the mentioned season that then the ice conditions were very differentiated in various water areas. The south and west Baltic ice severity was similar to the Gulf of Bothnia; in this area neither Finnish nor Swedish icebreakers operated. Moreover, in March, meteorological conditions in the Gulf of Bothnia proved to be highly changeable (drift ice in several hours displaced between the coastal zones of Sweden and Finland).

The correlation coefficient between the real number of icebreakers assistance events and the one calculated on the basis of the formula equals to 0,88.

4 "S" AND "K" INDICATORS DEPENDENCE ON CLIMATE CONDITIONS

Functional dependencies between "S" ice index and the regional driving factor (the data from re-analysis for the same climate conditions – the parameters and spatial location – as the greenhouse gases emission scenarios results and ECHAM model for the XXI century) were analyzed and processed for the calibration period of 1956-2004.

Primary, the influence of climate was represented by 129 parameters, affecting formation of sea ice in the Baltic Sea; these parameters were taken into account for the calibration period of 1956-2004. They were mainly the annual and monthly average air temperatures and their combinations, also the component indicators (meridian and zonal) of the atmospheric circulation (monthly, annual).

The next step of this work was to formulate the relationship (as the function) between ice severity indicator "S" and climate parameters. Construction of the optimal predictors set has been made on the basis of the carried out analysis results: correlation

(statistically significant at 95%; level: the highest value of the correlation coefficient between the predictors and response variable and the lowest between predictors), genetic algorithm (the longest "survival" of the predictors) and model sensitivity tests (better/worse work of the model with/without analyzed predictor). The researches proved that the sea-ice severity indicator "S" is the most sensitive to the mean monthly air temperatures in July and December and the mean annual air temperature as well as the zonal component of the atmospheric circulation. Therefore these parameters were used in constructing the models.

Relations between the parameters were calculated for the calibration period of 1955-2004 by multiplying linear regression for 3 types of the models.
- "thermal" (the dependence on the mean monthly temperature of July and December further called M1a model ;
- thermal "B" (function of the average annual air temperature and of the mean monthly December temperature)- further called M1b model
- c "thermal-zonal" (dependence on the mean air temperature of July and December and the zonal component of the atmospheric circulation) - further called M2 model

Comparison of the data from the period of observation (1956-2004) with the calculation results proved a statistic consistency, characterized by the following correlation coefficients: 0,6 (model 1a); 0,64 – model 1b and model 2.

To enable potential changes of navigation conditions in ice (represented with "K" indicator) in the XXI century there have been used combinations of formula 1 and processed dependencies between "S" and the climate parameters (items a, b and c)

5 RESULTS

To predict changes of the sea ice severity "S" indicator value in the XXI century the processed and presented in the section 3 models have been used; also to settle a number of assistance events "K" – formula1.

The climate parameters, which the models were based on, have been taken from calculations of the IMGW PIB Climate Deparments and they are an effect of adaptation of three global climate scenarios B, A1B and A2 run1 and also ECHAM–5 to the regional conditions.

It means that each of the models (M1a, M1b and M2) was calculated for 3 scenarios (B, A1B and A2 run1 and also ECHAM–5). To differentiate particular models and scenarios the following concept of designation was adopted:
- M1a Br1 – stands for the model based on the dependence "S" on the mean monthly temperature

of July and December) reckoned using the data taken from the scenario B run 1,

- M1a A1Br1 – the model based on the dependence "S" on the mean monthly temperature of July and December) calculated using the data taken from the scenario A1B run 1,
- M1a A2r1 – the model based on the dependence "S" on the mean monthly temperature of July and December) calculated using the data taken from the scenario A2 run 1,
- M1b Br1 is a model based on dependence between S and the average annual air temperature and of mean monthly December temperature calculated using the data from the scenario B run 1,
- M1b A1Br1 is a model based on dependence between S and the average annual air temperature and of mean monthly December temperature calculated using the data from the scenario A1B run 1,
- M1b A2r1 is a model based on dependence between S and the average annual air temperature and of mean monthly December temperature calculated using the data from the scenario A2 run 1,
- M2 Br1 designates a model based on dependence between S on the mean air temperature of July and December and the zonal component of atmospheric circulation, calculated using the data from the scenario B run 1,
- M2 A1B designates a model based on dependence between S on the mean air temperature of July and December and the zonal component of atmospheric circulation, calculated using the data from the scenario A1B run 1,
- M2 A2 designates a model based on dependence between S on the mean air temperature of July and December and the zonal component of atmospheric circulation, calculated using the data from the scenario A2 run 1,

The results of calculation carried out for these types of models (3 models for 3 different scenarios) for a period of next 20 years (2011-2030) are presented in Table 1 with comparison thereof with the values obtained for 20 years of 1971-1990 period.

Values of the following parameters: standard error, standard deviation and range, are significantly lower in 2011-2030 period than in the reference period. However, the mean, median and minimum values are in 2011-2030 period higher than in the reference period. The Kurtosis value for M2 A1Br1 model is very close to the value calculated for the reference period, as is the skewness, calculated on the basis of M1b A2r1 and M2B1 models. A maximum value calculated on the basis of M1b A2r1 model is close to the maximal value of the reference period (4,28 and 4,79).

It means that concentrations of „S" values around the mean value are similar for the both periods under consideration (model M2 A1Br1), as well as asymmetry of calculations of M1b A2 r1 and M2 B1 models are. The values of the mean, median and minimum are higher while the maximum value is comparable, what indicates that in coming years there are expected (according to the assumed scenarios) more winters with a small number of days with ice (but still with ice) than in the reference period.

The course of "S" indicator in period 2011-2030, calculated on the basis of scenario A2r1, together with comparison with the last climate reference period is shown on Figure 3.

Process variation of "S" indicator in the twenty first century was estimated on the basis of three models, each for three scenarios. Results of these calculations for scenario A1B are presented in Figure (4).

In the course of the ice severity indicator, calculated for the twenty first century using three scenarios, one can see the similar characteristics, recognized for period of 2011-2030.

The estimated number of icebreakers' assistance events was calculated using formula 1 for each of 3 models of "S" for 3 climate scenarios. Numbers of the estimated occurrence of obstruction for navigation in ice are shown in figures 5-7. Model 1a is represented by thick, black line, model 1b by drops and model 2 by thin dark line.

Table 1. Statistical parameters of „S" indicator for 2011-2030 and the reference period 1971-1990.

Period	1971-1990				2011	-2030				
Input Scenario	real	M1a Br1	M1b B1r1	M2 B1r1	M1a A1Br1	M1b A1Br1	M2 A1Br1	M1a A2r1	M1b A2r1	M2 A2r1
Mean	1,92	2,34	2,53	2,14	2,45	2,50	2,17	2,53	2,91	2,51
Median	1,52	2,26	2,48	2,13	2,54	2,59	2,14	2,54	2,83	2,62
Standard deviation	1,45	0,54	0,76	0,55	0,50	0,58	0,45	0,47	0,64	0,54
Kurtosis	-0,46	0,32	-0,76	1,18	-0,24	1,21	-0,43	0,68	0,14	-1,09
Skew/skewness	0,82	0,29	0,10	0,79	-0,45	-1,04	0,17	0,61	0,72	-0,27
Minimum	0,11	1,20	1,34	1,19	1,41	1,05	1,28	1,65	1,86	1,60
Maximum	4,79	3,53	3,78	3,60	3,35	3,43	3,01	3,60	4,28	3,35

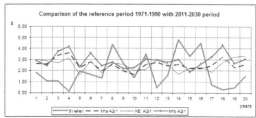

Figure 3. Comparison of the sea ice severity indicator calculated for the reference period and estimated for 2011-2030 period according to A2r1 scenario.

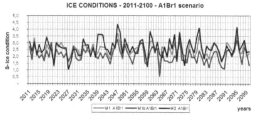

Figure 4. The course of the ice severity indicator in the twenty first century, calculated using A1Br1 scenario and three empirical models.

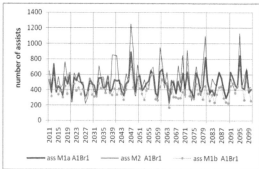

Figure 5. The changes of the estimated number of icebreakers' assistances in the twenty first century, calculated using B1 scenario and three empirical models (M1a B1r1 M1b B1r1, M2 B1r1)

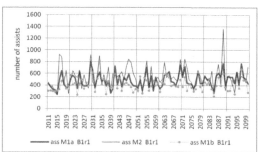

Figure 6. Changes of the estimated number of icebreakers' assistance events in the twenty first century, calculated using A1B scenario and three empirical models (M1aA1Br1, M1bA1Br1, M2A1Br1)

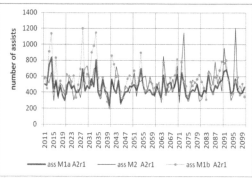

Figure 7. Changes of the estimated number of icebreakers' assistance events in the twenty first century, calculated using A2 scenario and three empirical models (M1a A2r1 M1bA2r1, M2A2r1)

According to all three scenarios, a number of icebreakers assistance cases should be lower than in the twentieth century. The lowest intensity of such decrease is estimated using model 2 and scenario A1Br1, the highest one – using model M1b and scenario B r1.

Otherwise, the minimum value, calculated for the scenarios, is higher than in 1956-2004 period. It means that probably the period, when obstructions for navigation in ice occur, could be longer, but not so severe as it happened in the period of 1956-2004. According to scenario B1r1 as well as the empirical model 2, intensity of such obstruction could increase during the 21 century. The same tendency is shown in scenario A1Br1 and using model 1a. The other models and scenarios have proved a decreasing trend up to 2100.

6 REMARKS

The aim of this work was investigation of changes in intensity of obstruction by ice occurrence (represented by the number of Swedish and Finnish icebreakers' assistance events, forced in effect of the climate change in coming 90 years of the twenty first century. The estimation was made under a hidden assumption that technical parameters of icebreakers and merchant ships would not change in the twenty first century, the same as intensity of Baltic navigation in ice.

It has been one of the first probes of technical application of the global climate scenarios changes. It should be stressed, that the presented work results (as application of climate scenarios), cannot be treated as a forecast, as they are only used to estimate the changes tendency.

ACKNOWLEDGEMENTS

This work has been achieved through Project KLIMAT Number POIG.01.03.01-14-011/08-00 ("*The impact of climate change on the environment, economy and society*") supported by the Programme Innovative Economy under the National Strategic Reference Framework, which is co-financed with EU resources.

The project is realized by the Institute of Meteorology and Water Management (Poland). The authors wish to thank prof. dr. Miroslaw Mietus (Institute of Meteorology and Water Management – National Research Institute) for valuable suggestions and participation in very useful discussion, also workers of IMGW PIB E.Jakusik MSc and R.Wójcik MSc for their support.

REFERENCES

Jakusik E. , Marosz M., Pilarski M., Mietus M, 2010 :Wpływ pola barycznego na wysokość falowania wiatrowego w południowej części Morza Bałtyckiego. Materiały Geo-Sympozjum Młodych badaczy Silesia, 1-14,

Jakusik E, Wójcik R., Biernacik D., Miętus M., 2010: Wpływ zmian pola barycznego nad Europą i Północnym Atlantykiem na zmiany średniego poziomu Morza Bałtyckiego w strefie polskiego wybrzeża , w Woda w badaniach geograficznych, UJK Kielce, 59-74

Kalnay et al., 1996, The NCEP/NCAR 40-year Reanalysis Project, Bull. Amer. Meteor. Soc., 77, 437-471

Mietus M., 1999, The role of atmospheric circulation over Europe and North Atlantic in forming climatic and oceanographic conditions in the Polish coastal zone, Materialy Badawcze IMGW, seria Meteorologia, 29, 157 pp. (in Polish),

Mietus M., (main editor), 2009: Report of KLIMAT project, www.klimat.imgw.pl,

Mietus M., ., (main editor) 2010 Report of KLIMAT project, www.klimat.imgw.pl,

Schmelzer N., Seina A., Lundqvist J.E., Sztobryn M., 2008, Ice, in: State and Evolution of the Baltic Sea, 1952-2005, Wiley & Sons, 199-240,

Storch von H., Zwiers F., 2001, Statistical analysis in climate research, Cambridge Univ. Press, 499 pp.

Sztobryn M.,2006, Sea ice condition – Polish coastal waters 1955-2005, IMGW, Report of Project DS.-H7, (in Polish)

Sztobryn M., Schmelzer N., Vainio J., Eriksson P.B., 2009. „Sea Ice Index". Report Series in Geophysics, University of Helsinki, No 61, 82-91, Helsinki 2009

Sztobryn M., R.Wójcik, 2010, Impact of climate change on the Baltic sea ice conditions, 20th IAHR International Symposium on Ice, Finland,

Werner P. C., Storch von H., 1993, Interannual variability of Central Europe mean temperature in January-February and its relation to large-scale circulation, Climate Research 3, 195-207

3. Storm-surges Indicator for the Polish Baltic Coast

I. Stanisławczyk
Institute of Meteorology and Water Management, Maritime Branch in Gdynia, Poland

ABSTRACT: Storm surges appear in the coastal zone of the Baltic Sea and, depending on row of factors, have different sizes, specifically characteristic for each region of the sea-coast. Observed climate changes are characterized with greater dynamics of weather phenomena. To compare the risk of storm surges to different areas, a new method had to be developed. Storm-surges indicator is used to compare the risks to the South Baltic water areas, varied along with conditions therein and the hydro meteorological and local conditions. The studies on the relations between the parameters and the occurrence of storm surges were carried out as well. The storm surges indicator "W" is related to the number of storm surges observed at the stations in the particular regions, the maximal wind velocity and the max sea level occurring during the same storm surges. The storm surges indicator was calculated for the period of 1955-2008 for the Polish coastal zone. The intention is to use this indicator for research and forecasting purposes. Assessment of the tendencies and variability of the regional phenomena indicators in timescale prove occurrence of certain regional changes of hydrometeorological conditions.

1 INTRODUCTION

Storm surges, as water level extremes, have been investigated quite extensively as they represent a major threat to the coastal population. Considerable technical and scientific effort has been invested worldwide to reduce the impacts of such phenomena, which may reach catastrophic proportions. Storm surges do occur within the Baltic coastal zone where, in addition to causing a direct threat to the seashore and inundation of the coastal area, they affect safety of navigation and operation of the ports. Climate change and sea-level rise will increase the frequency and severity of storm surges (based on the SRES scenarios in range 0.2-0,8 m/century). Observed climate changes are characterized with greater dynamics of weather phenomena. A number of weather and hydrological phenomena already now characterized by the increase in the prevalence of their presence and intensity of natural disasters.

Translocating storms induce surges, characteristic for specific seashore regions and, depending on a row of factors, the surges reach various sizes. To compare the risk of the storm surges at various water areas the new methods had to be worked out. Taking the above into consideration, there had been undertaken the works focused on classification of the storm surges sizes and developing a concept of the storm surge indicator, including analyzing the indi-cator values in relation to the South Baltic coast. The identification of the hydro-meteorological factors and the height of storm surges were done.

Thus the storm surge indicator is used to compare the storm surges related risks, occurring at various water areas. In the above studies there were analyzed the data gathered within a period of over 50 years, it means from 1955 to 2008 at stations of Świnoujście, Ustka and Hel. It was accepted (in accordance with the definition given by A. Majewski) that any hydrological situation, when the sea level is reaching or extending 570 cm, stands for a storm surge. (thus for Hel the alarm level was decided to be at 570 cm, for Ustka 600 cm and for Świnoujście - 580 cm).

2 METHODS

Sea level changes along the coasts are generated in connection with several factors, however mainly with the wind impact. Affecting sea surface by wind results in changes of the sea surface which appear in a form of storm surges of water. Within a time from November till March the largest number of storms used to occur; they generate extreme changes of sea levels. The most intensive deterioration of the seashore occurs in a time of severe storms. The sea coastal zone is an area covering both – on-shore part (coastal zone) as the off-shore one. The storm surges are threatening about 500 km of the coastline. The variable route of the coastline results in differentiated exposure of individual sections to wind conditions. In spite of considerable equalization of our coastline, there exist sections of different exposition to wind directions, characterized with diverse regime in meteorology and hydrodynamics.

Storm surges occur in the coastal zone of the Baltic Sea and, depending on a row of factors, have different sizes, characteristic for each region of the seacoast. Of essential influence are however, the local conditions. The magnitude and the character of the changes depend on the coast line configuration, on the bathymetry of the adjacent sea basin, on the exposition of a particular coast part to the actual wind etc. However the most spectacular deformation of the water surface at the Baltic Sea shores can occur due to the already mentioned off- or on-shore, stormy, sometimes hurricane-like winds.

Mostly active depressions with the fronts systems move eastwards from over the Atlantic Ocean. Well developed low-pressure troughs and their frontal systems moving across the coast are accompanied by gale-force backing winds as the fronts approach, and by veering winds after they have passed. The pressure gradient becomes very steep and the wind, initially gale-force, increases in severity and finally reaches hurricane force. In the forefield of a depression winds prevail of a strong southern component, behind the fronts usually veering. The winds in the forefield of a depression are offshore in relation to the southern coasts of the North and Baltic Seas. This situation normally causes sea level oscillations.

The mean sea level and a number of storm surges in the southern Baltic Sea have visibly increased during the last century. Figure 1 shows the number of storm surges in Świnoujście in 1950-2008, from 1950/1951 (from August 1950 to July 1951) to 2007/2008, when maximum occurrence of storm surges is for the autumn – winter months, from November to January – February.

The storm surges are not a regular annual event. Their number may vary as in the case of Świnoujście - from no occurrence at all to as many as twelve. The linear trend of this parameter values indicates an increasing tendency of the storm surges number (linear trend in Świnoujście is described as follows: $y = 1,6407+0,0619*x$). Also a threat of storm surges got increased towards the end of the 20 century nearly twice as compared to the middle of 20 century in the Southern Baltic Sea. Within 1950-1979 (almost half the entire period) 72 storm surges occurred, whereas in 1980-2008 as many as 129.

Distribution of the annual sea level maxima on the Polish coast (Świnoujście - on the western part and Gdańsk - on the eastern part of the Polish coast) changed also within 1955-2008. The annual sea level maxima frequency distribution (for 10 cm intervals) in Gdańsk (Fig. 2) in two periods - of 1955-1981 and of 1982-2008 illustrates the changes occurred in the period of issue.

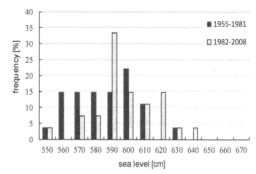

Figure 2. Frequency distribution of annual sea level maxima (for 10 cm intervals, e.g. 550-559, ect.), Gdańsk, 1955-1981 and 1982-2008

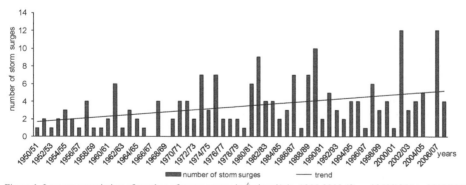

Figure 1. Long term variation of number of storm surges in Świnoujście, 1955-2008 (from 1950/1951 to 2007/2008).

The annual sea level maxima in Gdańsk varied from 570 cm (as in the definition - for Gdańsk the alarm level was 570 cm, although the lowest value of the annual sea level maxima is 557 cm) to 644 cm and the most frequent were values ranging between 590 and 600 cm in 1955-2008 (intervals are left-closed). In 1955-1981 period the most frequent were the peak values of the annual sea level maxima ranging between 600 and 610 cm (22%), then followed the values between 560 and 600 cm (the same values). The maximal values of the sea level ranged between 630 and 640 cm. In 1982-2008 the frequencies of the annual sea level maxima have been shifted to the higher values. However, the most frequent were the peak values between 590 and 600 cm, but the maximal annual sea level was in the higher range of 640-650 cm.

The frequency distribution of the annual sea level maxima (for 10 cm intervals) in Świnoujście within two periods - of 1955-1981 and of 1982-2008 is shown in the Figures 3. The distribution of the sea level maxima in Świnoujście differs from that of Gdańsk, but the frequencies of the annual sea level maxima in 1982-2008 haves been shifted also to the higher values.

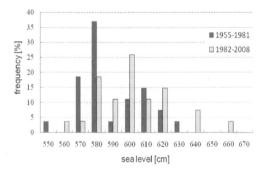

Figure 3. Frequency distribution of annual sea level maxima (for 10 cm intervals, e.g. 550-559 ect.), Świnoujście, 1955-1981 and 1982-2008

In Świnoujście the annual sea level maxima varied from 580 cm (as in the definition - for Świnoujście the alarm level was 580 cm) to 669 cm, and in 1955-2008, maximum was between 580-590 cm (intervals are left-closed). Thus the maximal annual values of sea level were shifted from a range of 630-640 cm to 660-670 cm. Distribution of the annual mean sea level (frequencies) indicates also an increase in their amount in a last period and trend of changes is also growing.

On base these of results one can ascertain, that in nearest period threat of storm surges on sea-coast will grow, what will influence on necessity of more precise forecasts and of information on theme of storm surges, flooding and coastal erosion hazard.

2.1 Storm surges indicator for the Polish coast

Proper classification is in terms of conditions of storm surges occurrence, gives the opportunity to compare and assess the risks of different and often separated areas. The storm surges indicator "W" indicator has been worked out to satisfy researches on climate, weather forecasting, navigation and sea ports operation planning. General aim is the development of innovative methods for mitigation of hazard in the context of increasing storminess and sea level rise. In addition, the research works were aimed at assessment of the storm surges threat to various water areas of the Polish Baltic coastal zone and correlativeness of the regional indicators and climate variability. The appropriate classification of regions in respect of surge occurrence conditions offers a possibility of comparison and assessment of the threat to differential, located often at quite a distance apart water areas and in future, a possibility of seasonal storm surges forecasting.

There has been preformed identification of factors, conditioning and determining storm surges occurrence at the Polish sea coast. The meteorological and hydrological factors have been indicated, analysis of the selected parameters carried out. There has been examined correlativeness of the maximal surge height and other parameters and magnitudes: atmospheric circulation, atmospheric pressure, wind velocity and direction, differences in water and air temperatures, also the mean monthly temperature of air within the specific years of the period in question in Świnoujście. The obtained considerable values of the correlation coefficients were used to select the parameters conditioning a storm surge size. On such a basis the concept of the storm surges indicator "W" has been worked out.

2.1.1 Storm surges indicator "W"

The indicator "W" is related to the number of storm surges observed at stations in the particular region and the maximal wind velocity and the maximal sea level occurring in a time of the same storm surge.

The storm surges indicator "W" comprises the parameters specified below; thus:

$$W = \frac{1}{i}\sum\left(\frac{V^2}{H \times 0.1}\right)_i$$

where i = number of observing storm surges in month, year or season; V = maximal wind velocity [m/s] observed within a time of a singular surge; H = maximal high of sea level [m] within a time of a singular surge; and 0.1 = numerical coefficient [m/s²].

Thus; the storm surges indicator "W" is used to compare a threat with storm to different areas. It also displays a potential risk which may occur in circumstances of changing the meteorological conditions –

a path of low, which is a random phenomenon, and, what is followed therewith, eventual wind force and direction changes to more unfavorable and more danger. The suggested formula of calculation of the indicator enables to assess estimatively a threat with high surges even basing on non-homogenous values and analyzed periods.

3 RESULTS

To estimate (potential risk) the threat of storm surges on the Polish sea coast, selected were the data from 3 sea areas, representatives of different kind of hydrographic conditions. In storm surges occurrence there can be seen regional differentiation, reflected in regional diversification of the storm surge indicator.

To determine the South Baltic storm surges indicator (seashore of Poland) there have been selected stations, representing dissimilar types of hydrographic conditions: Świnoujście (the Pomeranian Bay), Ustka (open waters of central seaside) and Hel (area of the open sea) In the analysis there were applied the homogeneous sequences of data collected by the mentioned stations within a period of over 50 years (1955 – 2008). Distribution of the indicator's values is different in specific regions of Polish sea coast. The results are presented in Table 1.

Table 1 Statistical parameters of the storm surges indicator "W" on the Polish coast, 1955-2008 and 1971-1990.

Area	Storm surges indicator					
	1955-2008			1971-1990		
	Max	Min	Mean	Max	Min	Mean
Świnoujście	7.4	0.2	2.8	6.3	0.8	3.1
Ustka	7.5	0.2	2.6	5.4	0.2	1.8
Hel	6.8	0.2	2.2	5.2	0.3	2.2

The values of the regional storm surges indicator range from 7.5 to 0.2, but in substance to zero (zero value proves that no storm surge occurred in a given time period). A number of years with no storm surge occurrence is as follows: within the mentioned period in Świnoujście - 5 years with no storm surges, in Ustka - 9 years and in Hel 12 years. The highest storm surges indicator "W" values within the whole period, i.e. 1955-2008, is related to Ustka (7,5), slightly lower in Świnoujście (7.4), and even in Hel (6.8), anyhow the mean indicator values are highest for Świnoujście (2.8), then for Ustka (2.6); the lowest for Hel (2.2). The storm surges indicator values for the period of 1971-1990 (period of reference is 30 years) for all the stations - Świnoujście, Ustka and Hel, were lower than the values for the whole period in question. Then, the mean values were higher in Świnoujście, but lower in Ustka, whereas for the Hel locality the mean value of the indicator was the same for the whole period of the investigations.

For Świnoujście the indicator value's standard deviation is 1.35, variance 1.8. The value of median, the average, it means such a value, of which 50% of surges were sever and 50% were mild is 2.7. The 90th percentile, it means a value, above which 10% of all the cases fall in, was equal to 4.8. For the Hel locality in turn, a value of the storm surges indicator is of standard deviation equal to 1.2, the variance is 1.4. The value of median is 1.8, the 90th percentile was equal to 3.9. For Ustka, the standard deviation is 1.5, the variance 2.2. The value of median is 2.4, 90th percentile was equal to 4.5.

The annual storm surges indicator within 1955-2008 period at the stations of Świnoujście, Ustka and Hel is presented respectively in Figures 4, 5 and 6. In Świnoujście the indicator is definitely of higher frequency, also its mean values are higher. The annual storm surges indicator changes within a range from zero to values above 5.

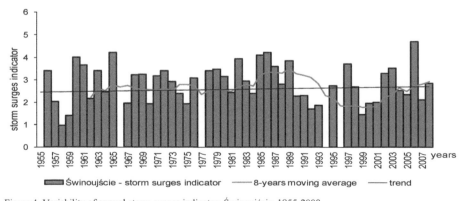

Figure 4. Variability of annual storm surges indicator, Świnoujście, 1955-2008.

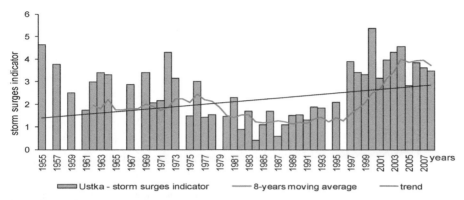

Figure 5. Variability of annual storm surges indicator, Ustka, 1955-2008.

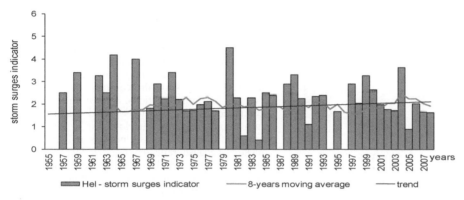

Figure 6. Variability of annual storm surges indicator, Hel, 1955-2008.

Values above 4 confirms frequent occurrence of high surges in storms of significant strength. A value close to 1 indicates a majority of storms, with slightly exceeded state of emergency. There also happen years with no high surges at all. A values of the indicator may change significantly even from year to year. The highest values of the annual storm surges indicator have been observed in Świnoujście in 2006, 1965 and 1986, in Ustka in 2000, 1955 and 2004 and in Hel in 1980, 1964 and 1967.

In Hel and Świnoujście localities there remains a stable value of the annual storm surge indicator (only in the nineties in Świnoujście there are of the considerably lower values), whereas in Ustka there occurred a high growth thereof since the middle of nineties, what confirms a course of anomalies of the annual storm surge indicator. The differences in the indicator variability course, fairly equal before (8 - years moving averages) is observed from the beginning of the eighties.

The indicator's value reflects not only the risk caused by surge, which occurs with very high sea level. It shows a possibility of much higher risk in case of wind direction changing to less favorable, also in case of a lower surge and very strong wind of not as much danger direction. For example: in Hel in 1999 the indicator value was 5.0, sea level was 580cm, the mean wind velocity 17m/s, but the wind direction only 260 grades, it means from the west directions. If only the wind direction had changed to less favorable, the north one, a potentiality of a very high surge would be much higher.

Anomalies of the storm surge indicator for Świnoujście, Ustka and Hel localities also confirm its regional character (Fig.7, 8 and 9). The results for various stations (areas) differ considerably from each other. Only for Ustka (in the central coast region) there is observed a clearly growing tendency in the storm surge course and the increased values thereof in recent years (approximately last ten years period).

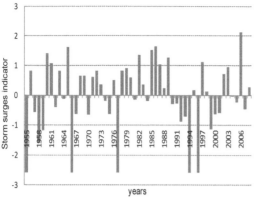

Figure 7. Anomalies of annual storm surges indicator „W_a", Świnoujście, 1955-2008 (mean W_a =2,58).

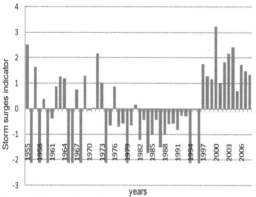

Figure 8. Anomalies of annual storm surges indicator „W_a", Ustka, 1955-2008 (mean W_a =2,12).

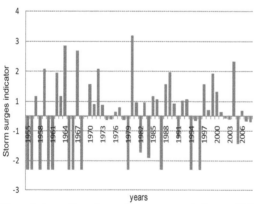

Figure 9 Anomalies of annual storm surges indicator „W_a", Hel, 1955-2008 (mean W_a=1,82).

Were analyzed the reasons for this increase, the differences between the periods 1955-2008 and 1997-2008 are connected with an increased mean wind velocity which occurred in Ustka in the last period, assuming the not changed wind directions dis-

tribution. A growth of the mean wind velocity is the most intense in case of the north directions, especially N (north) and NE (north-east) (Fig.10).

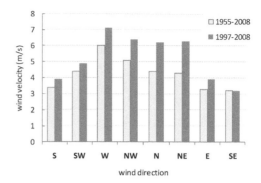

Figure 10. Frequency distribution of the mean wind velocity in Ustka, 1955-2008 and 1997-2008

This increase for mean wind velocity is 2 m/s for NE (north-east) direction and 1.8 m/s for N (north) direction. At the remaining stations are not observed any such changes, on the contrary, the mean wind velocity for north directions, is lower in recent years compared with the previous period: slight decrease in Świnoujście and greater lowering in Hel. The increase of the mean wind velocity is an important indicator, because reflected into an increase of the strong or stormy winds and higher storm surges. Assessment of the tendencies and variation in timescale of the regional phenomena indicators shows certain regional changes of climate conditions.

The seasonal storm surge indicators values enable assessment of its intensiveness within specific year seasons; it means autumn, winter and springtime, for the selected stations within a period of 1955-2008, assuming that autumn comprises months from September to November, winter – December to February and spring – March till May. In summertime the storm surges usually do not occur, however two cases happened in July 1989 and August 1995 in Świnoujście. Different seasons vary considerably in terms of value storm surge indicator for each part of the Polish coast. The seasonal storm surge indicator values are highest at all the stations in autumn and among them the highest value in Świnoujście - 3,1 (Table 2). Next in the order of the season, is winter and slightly lower values define the severity of the storm surges conditions during the spring season (but only in Świnoujście in springtime is higher indicator than in winter - 3,0).

Table 2. Seasonal storm surges indicator on the Polish coast, 1955-2008.

Area	Storm surges indicator		
	Autumn	Winter	Spring
Świnoujście	3.1	2.6	3.0
Ustka	2.8	2.6	2.4
Hel	2.5	2.1	2.1

A very large influence of local factors, resulting in large differences in storm surges conditions and value of storm surges indicator of each areas. In Świnoujście area storms conditions cause much more threat than in the other part of the coast.

4 CONCLUSIONS

To compare the risk of the storm surges at various sea areas, the new methods had to be developed. The storm surges indicator has been worked out to satisfy researches on climate, weather forecasting, navigation and sea ports operation planning. Weather and hydrological phenomena characterized now by the increase in their presence and intensity.The presented storm surge indicator allows for satisfactorlly characterising the conditions of threat with surges to Polish coastal zone and changes thereof in timescale. The comparative analysis of the indicator has confirmed its regional character. Moreover, the observed conditions variability in the last years proves that there occurs an increased threat with large surges in storms of considerable severity for a given area of the coast. In case of Świnoujście the threat is most intense however it remains at stable level. At the central area of the coast (Ustka) there can be observed a clearly growing tendency within a course of the storm surge indicator values (with its increased values in last ten years). The least threat occurs around Hel locality; within the many years course of the annual indicator there is to notice a slightly growing trend only and fairly low values thereof. Assessment of the tendencies and variability of the regional phenomena indicators in timescale prove occurrence of certain regional changes of climate. It is planned to carry out further work on the capabilities of the practical application of the indicator and the relationship between the storm surges indicator and regional processes and scenarios of changes (to 2030).

ACKNOWLEDGEMENTS

This work was supported by The Programme Innovative Economy under National Strategic Reference Framework, co-financed from EU resources. Project KLIMAT under grant No POIG.01.03.01-14-011/08-00 (The impact of climate change on the environment, economy and society)

REFERENCES

Lisitzin, E. 1974. Sea-level changes, Elsevier Scientific Publishing Company, Amsterdam – Oxford – New York

Majewski, A. 1961. General characteristic of sea level changes on the Southern Baltic Sea, Bull. PIHM 4, Warszawa

Majewski, A., Dziadziuszko, Z., Wiśniewska, A. 1983. The catalogue of storm surges on the Polish coasts in the years 1951-1975 (in Polish). Warszawa: Wyd. Kom. i Łączn.

SRES scenarios - The Special Report on Emissions Scenarios, prepared by the Intergovernmental Panel on Climate Change (IPCC), 2001,

Stanisławczyk, I. 2002. Meteorological conditions of storm surges on the southern Baltic Coast in period 1976-2000, Conf. NT Maritime Safety and Protect of Natural Environment, NOT Koszalin, pp 95-102,

Stanisławczyk I., Sztobryn M., Kowalska B., Mykita M., 2008, "Climate of Low Sea Levels on the Southern Baltic Sea Coast." Polish Journal of Environmental Studies, Olsztyn, 205-212,

Stanisławczyk I., Kowalska B., Mykita M., 2009. Low sea level occurrence of the southern Balic Sea coast. Monogr.: Marine Navigation and Safety of Sea Transportation – Weintrit A. (ed.), CRC Press/Balkema, 473-478, Leiden, Holandia

Sztobryn, M., Stigge, H-J., Wielbińska, D., Weidig, B. & Stanisławczyk, I. et al. 2005. Storm Surges in the Southern Baltic Sea (Western and Central Parts), Monogr., Bundesamtes für Seeschifffart und Hydrographie: 39, 51, 3, 51.

Sztobryn M., Weidig B., Stanisławczyk I., Holfort J., Kowalska B., Mykita M., Kańska A., Krzysztofik K., Perlet I.. 2009. Negative surges in the southern Baltic Sea (western and central parts). Berichte des Bundesamtes für Seeschifffahrt und Hydrographie Nr. 45/2009, 73

Wielbińska, D. 1964. Influence of atmospheric situation on sea level. Bull. PIHM 2.

4. Polish Seaports – Unfavorable Weather Conditions for Port Operation (Applying Methods of Complex Climatology for Data Formation to be Used by Seafaring)

J. Ferdynus
Gdynia Maritime University, Poland

ABSTRACT: The use of methods of complex climatology, which treats climate as a many-year weather regime, made it possible to define the annual structure of weather states observed in seven Polish sea ports, i.e. Elbląg, Gdańsk, Hel, Łeba, Ustka, Kołobrzeg and Świnoujście. This work uses data originating from OGIMET and covers the period 2000-2009.
Weather conditions that make port operation either difficult or even impossible are considered to be those when, during a day, we observe negative air temperatures, cloudiness, precipitation and strong winds at the same time. Once the frequency of occurrence of unfavorable weather conditions has been defined for the port operation then for each port a climatogram was drawn illustrating their frequency in the following decades of a given year.
Weather conditions which are unfavorable for the work in Polish sea ports are observed only in autumn and winter, and during early spring; they are most frequent in the ports of Ustka and Gdańsk. Their annual frequency in none of the described ports exceeds 1% so the conditions in Polish sea ports may be regarded as favorable for port operations. The worst weather conditions are observed in the last decade of December, third decade in January and in the second and third decades in February as well as i the first decade of April.

1 THE PROBLEM

Description of climatic conditions which are found in aids to navigation (Pilots, Routing Charts, and Pilot's Charts) make use of average values of meteorological elements. Such an approach follows the methods applied in classical climatology where climate is treated as "a mean state of atmosphere in a many- year period".

Climatic characteristics based on averaged courses of meteorological elements and on their extreme values, in some situations, seem to have restricted use as it does not give any information as to the correlations between them. It says nothing about the real state of the atmosphere, about the weather. Complex climatology provides such a possibility as it treats climate (many–year weather regime) as "an average structure and a sequence of weather". This approach makes it possible to describe, at the same time, a series of meteorological elements observed, i.e. real state of atmosphere (real weather).

The frequency of occurrence of **types of adverse or unfavorable weather conditions** is extremely important for sea ports, for their effective work. No matter what cargo is handled in a given port, strong and very strong winds are to be treated as weather conditions which disturb or even do not allow work-ing properly. In number of cases the routine work of a port is made difficult when the air temperature falls below zero and remains as such during the whole or part of the day (when the air temperature is around or below 0° C during the day). Loading and discharging of some cargoes is impossible or ceased during precipitation. Poor visibility or visibility restricted by fog or any type of precipitation may cause great problems in ports both during vessels maneuvering and while loading certain kinds of goods. Because precipitation is difficult to forecast and because rain or snow or other types of precipitation is connected with clouds, so the occurrence of certain clouds can provide us with information which is important for port operation.

Port efficiency decreases significantly if, even one, of the mentioned meteorological element is observed. Thus, negative air temperatures result in freezing of moist bulk cargoes, precipitation restricts visibility and makes loading or discharging of goods which are susceptible to moisture impossible and strong winds affect safe navigation and have influence on proper operation of cargo handling facilities. Greater problems arise when the meteorological elements are observed together. With the air temperature falling below 0°C and strong wind which cools the air at the same time cause that port operation is difficult. When minus temperatures are accompanied by strong wind and precipitation then icing of vessels and cargo handling facilities can be observed.

A lot of hydrological and meteorological phenomena both dangerous and favorable for ports and navigation take place as a result of concurrence defined by certain liminal values of meteorological elements. The presentation of hydro-meteorological data proposed by complex climatology can make the assessment of climatologic conditions for this type of work easier and can help to define the periods which are marked by adverse or favorable climatic conditions with reference to port operation.

2 METHOD AND OBSERVATIONAL DATA

Basic theory regarding methods of complex climatology as well as its detailed description can be found in work by Olszewski (1967) and Woś (1970, 1977a and b). This analysis makes use of partially modified weather classification proposed by Marsz (1992). This classification, as well as a detailed description of procedure used in dealing with input data can be found in author's earlier works (among others 1994, 1996, 1997), and the way of interpretation is presented by Ferdynus, Marsz and Styszyńska (1995) and Ferdynus (1996, 2000).

The classified period comprises a given day and the elements by which it is characterized - covered mean minimal and maximal air temperature (T), mean overall cloudiness (N), sum of atmospheric precipitation (R) and mean and maximum wind speed (V). In this way each day is described by means of four figures – TNRV and the number of possible weather conditions in such classification amounts to 486 (9×3×2×9) – see Table 1.

Fig. 1. Location of the meteorological stations used in his study.

The data used above were taken from 7 ports (Fig.1) from the ten-year period 2000 – 2010. Daily values of meteorological elements originate from OGIMET data sets. They are averaged values of daily synoptic observations. These data were thoroughly checked and in doubtful cases were compared to the data originating from ECA&D and where it turned to be necessary, they were corrected accordingly.

Table 1. Classification of weather's

Symbols		Partitions	Name of weather
T	0	$20,0° < t_{av} < 29,9°C$, $t_{min} \geq 0°C$	exceptionally warm
	9	$10,0° < t_{av} < 19,9°C$, $t_{min} \geq 0°C$	very warm
	8	$5,0° < t_{av} < 9,9°C$, $t_{min} \geq 0°C$	warm
	7	$0,0° < t_{av} < 4,9°C$ $t_{min} \geq 0°C$	moderately warm
	6	$t_{min} < 0°$ and $t_{max} > 0°$	transitional
	5	$-0,0° < t_{av} < -4,9°C$, $t_{min} < 0°C$	moderately frosty
	4	$-5,0° < t_{av} < -9,9°C$, $t_{min} < 0°C$	frosty
	3	$-10,0° < t_{av} < -19,9°C$, $t_{min} < 0°C$	very frosty
	2	$-20,0° < t_{av} < -29,9°C$, $t_{min} < 0°C$	exceptionally frosty
N	1	$0,0 < N < 2,0$	blue sky
	2	$2,1 < N < 5,9$	party clouded
	3	$6,0 < N < 8,0$	cloud
R	0	$RR = 00$ mm	no precipitation or precipitation < 0,1 mm
	1	$RR > 00$ mm	precipitation
V	0	$0,0 < v_{av} < 1,5$ m/s	calm or light air
	1	$1,6 < v_{av} < 7,9$ m/s, $v_{max} < 11$ m/s	light breeze
	2	$1,6 < v_{av} < 7,9$ m/s, $v_{max} \geq 11$ m/s	light breeze with periods of strong breeze
	3	$8,0 < v_{av} < 16,9$ m/s, $v_{max} < 17$ m/s	strong breeze
	4	$8,0 < v_{av} < 16,9$ m/s, $v_{max} \geq 17$ m/s	strong breeze with periods of gale
	5	$8,0 < v_{av} < 16,9$ m/s, $v_{max} \geq 30$ m/s	strong breeze with periods of storm
	6	$17,0 < v_{av} < 29,9$ m/s, $v_{max} < 30$ m/s	storm
	7	$17,0 < v_{av} < 29,9$ m/s, $v_{max} \geq 30$ m/s	storm with periods of hurricane
	8	$v_{av} \geq 30$ m/s	hurricane

With reference to their location the examined ports can be divided into those located in the region of the Vistula Lagoon (Zalew Wiślany), the Gulf of Gdańsk, those located on open sea and in the region of Szczecin Lagoon (Zalew Szczeciński) (Fig.1). When taking into account morphological features, we can distinguish the following types of ports: those situated in bays, gulfs, e.g. (Gdańsk – the North Port), those located close to river mouths (Łeba, Ustka, Kołobrzeg), and river and canal ports (Świnoujście, Elbląg, Gdańsk – Inland port), and those located on open sea (Hel). If we take into consideration their size (e.g. cargo handling capacity or the port area) then we can talk about big and small ports.

3 FREQUENCY OF ADVERSE WEATHER CONDITIONS FOR PORT OPERATION

As it was mentioned in the introduction, weather elements which have influence on the efficiency of port operation are the following: low air temperature, strong wind and clouds. It was also denoted that concurrence of these elements is extremely unfavorable. This work, making use of classification mentioned above (Table 1), treats weather conditions, which are regarded as those having negative influence on port operation, as:

$T \cap N \cap R \cap V$ – group A;
$(T \cap R \cap V) \cup (T \cap N \cap R)$ – group B;
$(T \cap R) \cup (T \cap V) \cup (R \cap V) \cup (N \cap R)$ – group C;

where

$T = 2, 3, 4, 5, 6$;
$N = 3$;
$R = 1$;
$V = 3, 4, 5, 6, 7, 8$.

Weather types from Group A should be treated as especially unfavorable for port operation as they combine concurrence of all four adverse weather elements. Group B is made up of three of these elements and in Group C there are two meteorological

elements which make port operation difficult and in this way lower the port efficiency. Weather conditions which do not fall into any of the above groups are regarded to be neutral and are classed as Group D.

The examined ports are situated in the same climatic zone (maximum difference in latitude is only about 1°φ), so the observed differences in the frequency of occurrence of certain weather groups can only be attributed to local conditions.

3.1 Elbląg

The port of Elbląg (φ = 54°10'N; λ = 019°23'E) is the biggest port located on the Vistula Lagoon, situated on the river Elbląg in a distance of 6 km from its mouth to the Vistula Lagoon. It is a regional port rendering services for coastal navigation both for merchant (coal, building materials, sand, broken stone) and passengers vessels and for tourists. In 2008 only 14 vessels of total tonnage 4.6 thousand GRT called at the port of Elbląg (ten times fewer than in 2005). In 2008 cargo handling reached 5700 tons, in this amount 1700 tons of steel constructions and 4000 tons of sand.

The transport of passengers amounted to 39909 people. In 2008 cargo handling and passenger transport in the port of Elbląg reached totally 4000 tons and 32899 people respectively (Rocznik Statystyczny Gospodarki Morskiej 2009).

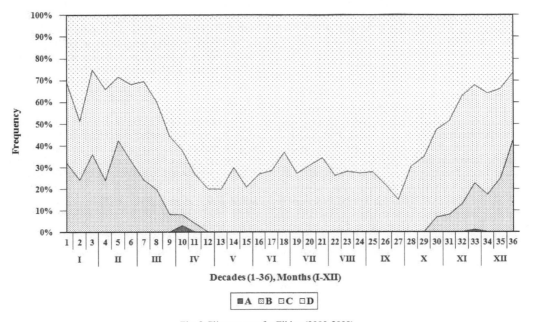

Fig. 2 Climatogram for Elbląg (2000-2009).

The analysis of climatogram drawn for Elbląg (Fig. 2) indicates that in the years 2000 – 2009 the type of weather from Group A is observed very seldom, accounting to only 1.0% in the first decade of December and 3.0% in the first decade of April – and they are the only two decades during the whole year. During the first decade of April there were two weather types 5314 and 6314 and in the first decade of December weather 6314 type.

If we take into consideration the occurrence of weather types from Groups B and C it is obvious that they are much more frequent. Weather types from Group B can be observed for the first time in the third decade of October and the last time in the first decade of April. They are noted in the third decade of December most frequently (around 50%). In the second decade of February a secondary maximum of frequency 42% is observed and from this decade on we can observe the decrease in their frequency. Weather types from Group C are noted in Elbląg during the entire year and they occur most frequently during winter and on the turn of seasons (second decade of November – 50%; first decade of December – 46%; first decade of March).

When we analyze the climatogram it is easy to notice that the weather type from group D comprise the rest of the dominant weather and it means that climatic conditions in the port of Elbląg seem to be favorable. From the third decade of March on, the frequency of that group exceeds 50% and such situation is observed till the third decade of October. The most favorable conditions for the port operation in the port of Elbląg are noted from the second decade of April to the third decade of September – during that period the frequency exceeds 70% and in the third decade of May reaches 87% and in the first decade of May 80%.

3.2 Gdańsk

In the port of Gdańsk ($\varphi = 54°24'N$; $\lambda = 018°42'E$) located in the central part of the southern coast of the Baltic Sea we can distinguish two regions of different operational parameters: Inner Port located along the Martwa Wisła and the North Port directly accessible from the Gulf of Gdańsk.

In the Inner Port there are container terminal, ferry terminal and terminal for ro-ro vessels, car terminal, and terminal for handling citrus, liquid and granulated sulphur and for phosphates. The remaining quays have universal character and make it possible to handle general cargo and bulk cargo. Terminals for handling fuel oil, liquid fuel and coal are located in the North Port.

In 2008 the port of Gdańsk served 3999 vessels of total capacity 32793,3 thousand GRT. About 19

mln ton were handled including ore – 16 thousand tons, grain 960 thousand tons, general cargo 3,5 mln tons, coal 2,7 mln ton and fuel 9 mln ton and almost 165 thousand passengers were served (Rocznik Statystyczny Gospodarki Morskiej 2009).

It can be easily noticed when comparing climatogram from Gdansk with the one for Elbląg that weather types from Group A are more frequently noted in the port of Gdańsk. The first record is observed in the second decade of December and it is observed till the first decade of April and their maximum frequency is noted in the second decade of February (at an average six days in a decade). They are weather types 4313, 5313, 5314, 6313, 6314 and 6315. When compared with Elbląg they are characterized by lower air temperature and greater wind speed.

Weather from Group B in Gdańsk is observed from the second decade of October to the first decade of April. Maximum, exceeding 20% frequency is noted in the third decade of December (24%), in the first decade of January (21%) and second decade of February (24%). Weather types from group C are noted during the whole year-in two decades they reach 50% frequency – the third decade of January and November.

When we analyze the frequency of weather which is favorable for the port operation i.e. group D, it should be noted that only in 10 decades their frequency does not exceed 50% (the lowest frequency 25% - in the second decade of February). In 14 decades their frequency exceeds 75% – such situation can be observed from the second decade of April till the third decade of September. The best weather conditions for port operation in Gdańsk occur in the third decade of May (91%). Generally speaking similar to Elbląg weather conditions in Gdańsk are favorable for port operation but they are a bit more difficult than in Elbląg.

3.3 Hel

The port of Hel ($\varphi = 54°36'N$, $\lambda = 18°48'E$) is located in the southern part of the Hel Sandbank, at the western part of the extreme part of the sandbank, on the eastern coast of the Puck Bay. The civil port is situated in the south west part of the town, south of Hel military port. The port of Hel is mainly fishing harbor and is also a port for pleasure crafts (White Fleet) coming from "3city". Hel has never been a cargo handling port (Krośnicka 2007). In 2008 Hel was visited by 1066 vessels of total capacity 479,8 thousand GRT (Rocznik Statystyczny Gospodarki Morskiej 2009).

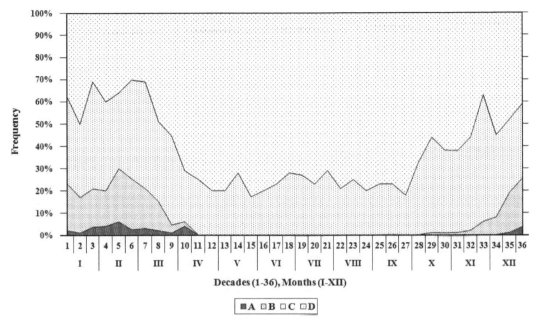

Fig. 3. Climatogram for Gdańsk (2000 – 2009).

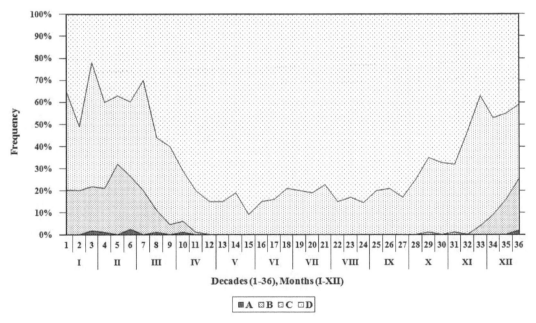

Fig. 4. Climatogram for Hel (2000 – 2009).

Weather types from Group A were observed in Hel only in 6 decades (in Gdańsk in 12 decades). Reaching maximum 2% frequency in the third decade of December, January and February - these weather types are marked with symbols 4314, 5314, and 6314 (Fig. 4). Contrary to Gdańsk the weather types from Group A occur in Hel in single, separated decades.

In the analyzed period 2000 – 2009 weather types form Group B appeared for the first time in the second decade of October, then in the first decade of November and only from the third decade of November they are constantly observed. Such situation is present until the second decade of April. Starting from the last decade of December till the first decade of March weather types form this group reach 20% frequency (maximum 32% in the second decade of February). Weather types form Group C are observed during the entire year and their maximum frequency in noted in the third decade of January (63%) and November (59%) and the first decade of March (50%).

Weather types from Group D are noted during the whole year and from the second decade of April till the first decade of October they reach 75% frequency- at the turn of May and June such weather conditions are noted 9 days in a decade on average.

3.4 Łeba

The port of Łeba (φ = 54°46'N, λ = 17°33'E) is situated on the mouth of the Łeba river (the place where the river flows into the sea) and similar to Hel is only fishing and tourist port. It is not fitted with cargo handling facilities (Krośnicka 2007).

Weather types form Group A (5313, 5314, 6313 and 6314) are observed for the first time in the second decade of November and for the last time in the first decade of April (Fig. 5). Similar to the port of Hel weather types from this group do not form one group; the decades during which they can be observed are separated and their frequency drops to 0%. All in all they are noted in 9 decades. When compared with the port of Hel and Gdańsk the frequency of weather types from Group A in Łeba is higher – they can be observed not only more frequently but also sooner, as early as in the second decade of November.

Weather types from Group B are noted from the third decade of October till second decade of April and the frequency exceeds 10% from the third decade of November till the second decade of March. They reach their maximum value during the third decade of December (37%) and January (35%) and in the second decade of February (31%). Weather types from Groups C are observed during the whole year and they reach maximum frequency (above 50%) from the second decade of October till the first decade of March; except the third decade in December, the second in January and the third decade in February.

The most favorable weather conditions for the port operation in the port of Łeba occur from the second decade of April till the second decade of June and from the third decade of July to the first decade of September, when the weather types from Group D reach maximum 75% frequency. When compared with the port of Hel weather types from this group appear more rarely.

3.5 Ustka

The port of Ustka (φ = 54°35'N; λ = 016°52'E) is located on the mouth of the Słupia river and is one of the largest fishing harbors in Poland-about 60 fishing boats permanently berth here. From a few to a dozen of thousands of tons of cargo is handled in the port of Ustka every year (in 2008 – 3.2 thousand tons). Most frequently it is bulk and general cargo. In 2008 twenty one vessels entered the port of Ustka (2 general cargo vessels, 11 barges carrying dry bulk cargo and 7 passenger vessels) of overall capacity 13, 6 thousand GRT (Rocznik Statystyczny Gospodarki Morskiej 2009).

The analysis of climatogram (Fig. 6) indicates that weather types from Group A (5313, 5314, 5316, 6313, 6314) occur in Ustka more frequently than in the previously described ports, reaching maximum 6% frequency in the second decade of February. For the first time they are observed in the third decade of November and for the last time in the first decade of April.

Weather types from Group B are observed from the first decade of October to the first decade of April and the frequency is more than 10% from the first decade of December to the second decade of March. The frequency exceeding 20% is noted in the third decade of December, in the first and third decade of January and first and second decade of February. Weather types from Group C are noted during the whole year - they exceed 50% frequency from the second decade of October to the first decade of January, in the third decade of January from the third of January to the second decade of February and first decade of March.

Weather types from Group D have 70% frequency from the third decade of April to the third decade of July, in third decade of August and September. The period of continuous occurrence of weather types from Group D is reduced when compared to the above described ports. The optimum conditions for port operations in Ustka are observed in the third decade of July (84%).

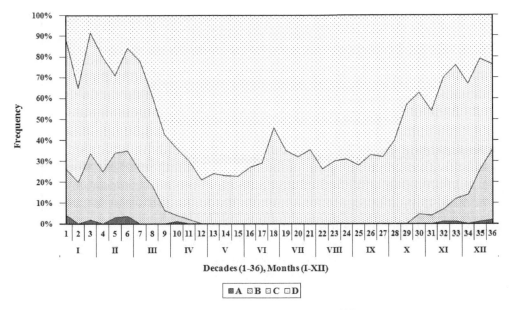

Fig. 5. Climatogram for Łeba (2000 – 2009).

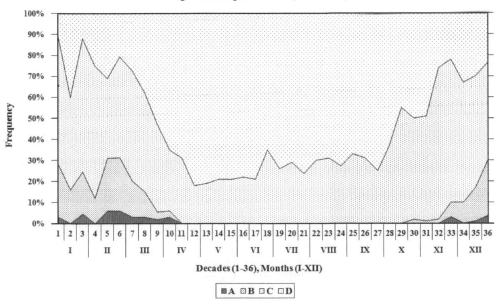

Fig. 6. Climatogram for Ustka (2000 – 2009).

39

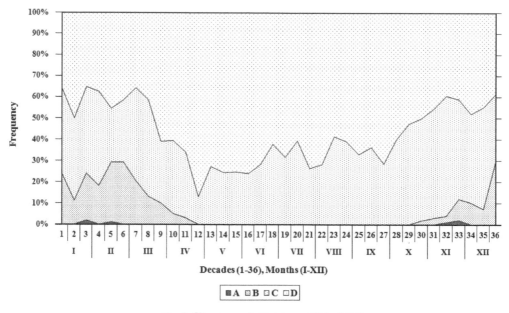

Fig. 7. Climatogram for Kołobrzeg (2000 – 2009).

3.6 Kołobrzeg

The port of Kołobrzeg (φ = 54°11'N; λ = 015°35'E) is situated on the mouth of the river Parsęta and is the sixth biggest merchant sea port in Poland. It handles about 150 thousand tons of cargo every year (84 vessels of total capacity 108,5 thousand tons in 2008). Bulk and general cargo prevail in the cargo handled there. About 60 fishing boats berth in Kołobrzeg and rendering services for them is one of the basic tasks of this port. Kołobrzeg is also tourist port which was visited by 93 passenger vessels in 2008 (Rocznik Statystyczny Gospodarki Morskiej 2009).

Weather types from Group A (6313,6314) appear in single decades, i.e. in the first and second decade of November, the third decade of January, and second decade of February; reaching maximum frequency of 2% (Fig. 7). The noted weather types are: 6313 and 6314. When compared with the previously described ports they are only weather types of periodical weather.

Weather types from Group B appear as early as in the third decade of October and remain until the second decade of April. The highest frequency is observed in the third decade of December (32%). In January and February, at the beginning of March their frequency is similar – reaching 23% and 24%. Weather types from Group C reach their highest frequency at the turn of October and November (more than 50%).

In the last decade of April and May weather types form Group B occur with the frequency of 80%. In a few decades they reach 70% frequency: it is in the first and the second decade of May, second decade of June and the first decade in July and third in September. The frequency of these weather types is not lower than 34% in any of the decades.

3.7 Świnoujście

The port of Świnoujście (φ = 53°54'N; λ = 014°15'E) is located on Wolin and Uznam islands, on the mouth of the Świna river. In 2008 the port of Świnoujście was visited by 5238 vessels of gross tonnage 63104,4 thousand tons. More than half of the cargo handled in the port of Świnoujście is made up of dry bulk cargo, with the majority of coal and coke. About 30% of all cargo handled are ro-ro cargo and lorries (Rocznik Statystyczny Gospodarki Morskiej 2009).

Weather types from Group A (6313) may be observed in the port of Świnoujście only in few separate decades and for the first time they are observed in the second decade of December (Fig.8). Apart from this case they are also noted in the second and third decade of February and March. They do not reach frequency higher than 2% in any of the above mentioned decades.

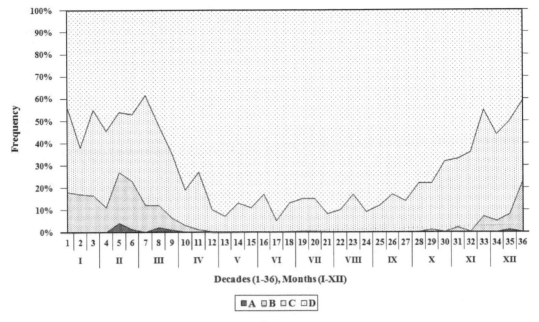

Fig.8. Climatogram for Świnoujście (2000 – 2009).

Weather types from Group B can be observed as early as in the second decade of October and in the first of November but continuously only from the third decade of November till the second decade of April, reaching 20% frequency only in two decades, i.e. in the third decade of December and second decade of February. The weather types from Group C never reach frequency of 50%.

Weather types from the last group reach maximum exceeding 90% frequency in a few decades-the third decade of April, May, July and August and the first decade of May and second decade in June. None of these are marked by frequency lower than 40%.

4 CONCLUSIONS

The analysis of climatograms drawn for Polish sea ports indicates that each of them has its own clearly defined structure of weather. In all of them both frequency as well as distribution in time of weather which is unfavorable for their operation is unique.

The weather types from Group A are most frequently noted in the port of Ustka and Gdańsk and least frequently in Elblag and Hel. In a case of Gdansk the weather of this group was observed in 12 and in Ustka in 11 decades, in Elbląg in two decades in Kołobrzeg in four decades, and in Świnoujście in five. Weather types from this group in none of these ports and in none of the decades reached frequency exceeding 10% which means that they are not ob-

served every year – so they are weather types occurring from time to time only. The weather types from Group A for the first time in an annual cycle appear as early as second decade of November (Łeba, Kołobrzeg) and the latest in the third decade of December (Hel). For the last time these types of weather are noted in the first decade of April (such a situation takes place in five out of seven analyzed ports). Thus, it may be stated that the most unfavorable weather types are observed, not surprisingly, only during autumn, winter and spring seasons. In the analyzed ports it is only Gdansk were the weather is noted in a continuous way from the second decade of December until the first decade of April. In the other ports decades with weather types from Group A are separated by decades during which such weather type is not observed.

Weather types which do not have direct influence on the efficiency of port operation, i.e. weather types characterized by positive air temperatures during the whole day, without clouds, without precipitation and calm are most frequently observed from April till September. The 80% frequency is observed in the decades starting from the first decade of April in Świnoujście, second decade in April in Hel, the third decade in April in Gdansk, Ustka and in Kołobrzeg, and in Łeba in the third decade in May. For the last time such high frequency is noted in the third decade of September (Elblag, Gdańsk, Hel, Świnoujście), in the third decade of August (Ustka) and in the third decade of July (Kołobrzeg). Thus, the periods covering summer, late spring and early autumn turn to be most favorable for carrying out port operations in Polish ports.

REFERENCES

Ferdynus J. 1994, Sezonowość klimatyczna Björnöyi w świetle rocznej struktury pogody. [w:] Problemy Klimatologii Polarnej 4, Gdynia, s. 119–138.

Ferdynus J. 1996, Struktura pogód w Tromsö [w:] Prace Wydziału Nawigacyjnego WSM z. 2. Gdynia, s. 84–93.

Ferdynus J. 1997, Główne cechy klimatu morskiego strefy subpolarnej Północnego Atlantyku w świetle struktury stanów pogód. WSM Gdynia, ss.138.

Ferdynus J., Marsz A. A., Styszyńska A. 1995, Możliwości wykorzystywania metod klimatologii kompleksowej do tworzenia informacji dla potrzeb żeglugi. VI Międzynarodowa Konferencja Naukowo – Techniczna Inżynierii Ruchu Morskiego. WSM. Szczecin, s. 93–103.

Marsz A. A. 1992, Struktura pogód i roczna sezonowość klimatu Stacji Arctowskiego.[w:] Problemy Klimatologii Polarnej 2. Gdynia, s. 30–49.

Rocznik Statystyczny Gospodarki Morskiej 2009, Główny Urząd Statystyczny.

Krośnicka K. 2007, Zagospodarowanie małych portów polskiego wybrzeża, ich stan techniczny, organizacja funkcjonalna oraz powiązania z otoczeniem. [w:] Praca zbiorowa pod red. A. Grzelakowskiego i K. Krośnickiej, Małe porty polskiego wybrzeża uwarunkowania i perspektywy rozwoju. Wydawnictwo Akademii Morskiej w Gdyni, Gdynia, s. 25–90.

Olszewski J.L. 1967, O kompleksowej charakterystyce klimatu. [w:] Przegląd Geograficzny t. XXXIX. z. 3, s. 601–613.

Woś A. 1970, Zarys klimatu Polski Północno-Zachodniej w pogodach. PTPN. Prace Komisji Geograficzno – Geologicznej. T. X. z 3. Poznań, ss. 156.

Woś A. 1977a, Klimatyczne sezony roku w Kaliszu. [w:] Badania Fizjograficzne nad Polską Zach. t. XXX. ser. A Geografia Fizyczna, s. 93–115.

Woś A. 1977b, Zarys struktury sezonowej klimatu Niziny Wielkopolskiej i Pojezierza Pomorskiego. UAM. Seria Geograficzna. nr 15. Poznań, ss. 90

5. Analysis of Hydrometeorological Characteristics in Port of Kulevi Zone

A. Gegenava & G. Khaidarov
Batumi State Maritime Academy, Batumi, Georgia

ABSTRACT: The presented paper continues paper titled "New Black Sea Terminal of Port Kulevi and it Navigating Features", which was published in "8th International Navigational Symposium on Marine Navigation and Safety of Sea Transportation. Gdynia, Poland, 2009" and deals with the aspects of safety navigation provision in Port of Kulevi by means of conduction of complex analysis of hydrometeorological characteristics in Port of Kulevi zone. As it was said in the previous paper, the port of Kulevi in the point of view of hydrometeorological conditions is a difficult one. The influence of prevailing wind directions – East and West, constant sea and the river Khobi currents, as well as the absence of the special protective hydrotechnical constructions allows the waves and sediments from the rivers Rioni and Khobi to enter the channel. The presented paper presents the analysis of hydrometeorological conditions for the creation of technical decisions whlch can be conducted in Port of Kulevi having the aim the decreasing of hydrometeorological conditions to provide the safety navigation.

1 INTRODUCTION

The fact that the systematic observations of hydrometeorological data in Kulevi zone were stopped in 1991 and all the data was taken from Georgia away, the taking into force of Black Sea Terminal in Port of Kulevi caused the necessity of constant monitoring of meteorological characteristics with the aim of safety navigation provision. The analysis of hydrometeorological characteristics shows that the main factor, which influences upon the level of safety of navigation in water area of Port of Kulevi is heavy sea. The analysis of wave conditions was conducted for the provision of the necessary level of safety of navigation as well as the provision of effective and safe maintenance of hydrotechnical constructions. The automatical meteorological station which is a part of Port of Kulevi VTS equipment gave possibility of every day (per each 30 minutes) keeping of the wind condition (direction, speed, maximal and minimal indicators). The complex of measures of Port of Kulevi water area protection was worked out on the basis of that analysis.

2 ANALYSIS OF METEOROLOGICAL CONDITIONS

Winds. Collection and analysis of the wind condition (direction, speed, maximal and minimal indicators) in Port of Kulevi zone was conducted by means of automatic meteorological station which is a part of Port of Kulevi VTS equipment. Data on the wind condition (each 24 hours, weekly, monthly, 6-monthly and annual) were worked out, calculated on average and rebuilt into the special graphs._By means of the graphs we tried to determine the prevailing direction, maximal and minimal speed of the wind in different period of time (See Fig.1.2.3).

Collection and working out data on the wind condition we started on 16.07.08, the presented analysis covers the period from 16.07.08 to 31.08.09, which is also divided into the following period of time (See Tab.1.): The analysis resulted in the following conclusions:

- Wind directions is distributed in the following way (See Tab.2. 3.4.5.6.7.8).
- Prevailing directions of the wind, as it can be seen in the graphs are the following (See Fig.1.2.3): ESE, East, WNW and West.
- Putting the axis of the channel on the "Wind Rose" it is shown that the prevailing directions of the wind and the axis of the channel are the same, allowing the safe entrance and leaving of the ships.
- The average speed of the wind is mainly between min – 2.2 м/s, max – 9.9 м/s and it is acceptable for safe entrance and leaving and quite lower than

criteria of safety safe maximal wind speed – 15 m/s, noted in Port Regulations.
– Maximal indicators of the wind force more than 15 м/s, are fixed in (See Tab.9.):
– The duration of the maximal indicators of the wind is between 4 – 6 – 8 hours. Than, with the change of the wind direction the speed decreases to the average indicators of – 3. 5 м/s – 5.0 м/s. The longer wind is detected from West, ESE, and East. They may blow during 2-3 days, sometimes for 5 days, in the period of June-October.

Table 1. Period of time collection and working out data on the wind condition

I Half-Year A	II Half-Year B	Annual C	Summer	Autumn D	Winter	Spring	Monthly E	Weekly F
16.07.08 – 31.12.08	01.01.09 – 31.08.09	16.07.08 – 31.08.09	16.07.08 – 31.08.08	01.09.08 – 0.11.08	01.12.08 – 28.02.09	01.03.09 – 31.05.09	-------	-------

Table 2. Distributed wind directions

	I Half-Year – A - 16.07.08 – - 31.12.08								
Wind Direction	ESE	E	SSE	WNW	W	WSW	SSW	ENE	S
%	31.7	17.0	8.5	8.4	7.2	6.8	5.4	4.6	4.1

Table 3. Distributed wind directions

	II Half-Year - B - 01.01.09 – 31.08.09								
Wind Direction	ESE	E	WNW	WSW	SSW	W	SSE	NNW	S	ENE
%	23.6	16.8	10.0	9.1	8.5	8.0	6.9	5.6	4.5	3.8

Table 4. Distributed wind directions

	Annual – C - 16.07.08 – 31.07.09								
Wind Direction	ESE	E	WNW	WSW	SSW	W	SSE	NNW	S	ENE
%	27.1	16.9	9.3	8.1	7.2	7.6	7.6	4.8	4.3	4.1

Table 5. Distributed wind directions

	Winter – D - 01.12.08 – 28.02.09					
Wind Direction	ESE	E	SSE	WNW	W	NNW
%	39.2	24.0	6.7	6.2	4.8	3.4

Table 6. Distributed wind directions

	Spring – D - 01.03.09 – 31.05.09								
Wind Direction	E	ESE	WNW	WSW	SSW	W	NNW	SSE	ENE	S
%	17.0	16.7	11.4	11.3	9.6	8.6	8.5	5.2	4.2	3.8

Table 7. Distributed wind directions

Summer – D - 01.06.09 – 31.08.09								
Wind Direction	ESE	WNW	WSW	W	SSE	E	SSW	S	NNW
%	22.5	11.7	11.3	11.2	10.0	10.0	8.9	5.3	4.2

Table 8. Distributed wind directions

Monthly – E, Weekly – F

Y	M	W	*	**	***	%
2008	7	3	3,91	9,20	319	30,26
2008	7	4	5,05	12,40	205	29,15
2008	8	1	4,94	10,90	207	31,10
2008	8	2	4,44	14,60	146	27,86
2008	8	3	3,75	12,60	84	41,04
2008	8	4	3,94	10,70	237	47,92
2008	9	1	4,16	7,40	296	77,89
2008	9	2	4,37	18,30	207	44,35
2008	9	3	5,31	18,00	106	47,92
2008	9	4	5,60	19,80	189	47,40
2008	10	1	8,09	28,40	195	41,87
2008	10	2	4,11	12,80	201	60,85
2008	10	3	4,48	12,30	103	45,83
2008	10	4	5,75	18,10	109	66,46
2008	11	1	4,42	12,30	296	55,95
2008	11	2	9,89	19,90	103	77,91
2008	11	3	4,49	11,00	102	71,14
2008	11	4	4,59	15,20	332	55,32
2008	12	1	5,79	16,60	109	66,77
2008	12	2	5,21	15,40	304	56,85
2008	12	3	8,51	15,90	103	77,38
2008	12	4	5,46	16,80	273	60,57
2009	1	1	5,93	19,20	103	76,42
2009	1	2	5,93	24,80	319	48,81
2009	1	3	4,58	11,90	255	57,44
2009	1	4	8,92	27,20	104	81,84
2009	2	1	5,64	18,30	102	57,44
2009	2	2	7,15	16,60	104	73,51
2009	2	3	7,36	22,90	103	60,90
2009	2	4	7,21	16,60	108	50,60
2009	3	1	7,69	14,50	99	68,15
2009	3	2	7,15	18,40	177	48,36
2009	3	3	6,81	18,40	101	43,45
2009	3	4	5,69	15,70	205	45,21
2009	4	1	5,20	17,70	105	34,82
2009	4	2	6,23	17,30	104	47,02
2009	4	3	4,72	19,80	104	27,98
2009	4	4	5,53	20,30	100	35,88
2009	5	1	5,21	13,20	98	21,43
2009	5	2	4,04	11,90	311	55,65
2009	5	3	3,99	12,50	245	36,90
2009	6	1	4,56	15,00	245	34,58
2009	6	2	5,73	17,80	302	44,94
2009	6	2	5,00	23,10	209	35,22
2009	6	3	6,38	21,10	102	25,60
2009	6	4	4,08	20,80	235	27,78
2009	7	1	4,30	13,30	297	47,02
2009	7	2	4,36	17,70	186	40,48
2009	7	3	4,38	14,20	185	36,61
2009	7	4	5,11	14,90	74	40,92
2009	8	1	3,76	10,90	246	63,39
2009	8	2	5,10	13,90	240	45,24
2009	8	3	4,54	18,00	353	49,70
2009	8	4	4,29	11,00	215	50,42
2009	9	1	2,24	2,90	106	85,71

*) average speed (from the maximal); **) maximal speed (from the maximal); ***) direction in the maximal wind.

Table 9. Maximal indicators of the wind force

Indicators	2008 Year			
	September	October	November	December
Wind (м/s)	18.3– 19.8	18.1–28.4	15.2–19.9	15.4–16.8
Direction	106^0; 207^0; 189^0	109^0; 195	332^0; 103^0	304^0; 273^0; 109^0

	2009 Year						
	January	February	March	April	June	July	August
Wind (м/s)	19.2–27.2	16.6–22.9	15.7–18.4	17.3– 20.3	17.8– 23.1	17.7	18.0
Direction	319^0; 103^0	103^0	205^0; 177^0; 101^0	105^0	102^0; 235^0; 302^0	186^0	353^0

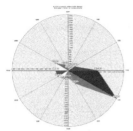

Figure 1. Prevailing direction of the wind in I Half-Year - A - 16.07.08 – - 31.12.08

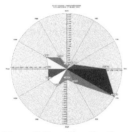

Figure 2. Prevailing direction of the wind in II Half-Year - B - 01.01.09 – 21.08.09

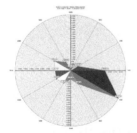

Figure 3. Prevailing direction of the wind in Annual – C - 16.07.08 – 31.07.09

Storms. The most stormy period is from the period from November to March. The most stormy month is November (average number of storms – 7, maximal – 14). The least number of the storms is in the summer – July-August. The bigger duration of the storm is in November and December and is 5 days (117 hours). There approximately are 15-17 days with the wind, which stormy speed and 15-17 days, when the wind speed is not less than 15-17 m/s. The most often stormy winds are from the East

– they repeat in more than 70%. The winds of wave-dangerous direction are the wind of western quarter and South-West. Stormy West and South-West winds repeat in 13-15%; strong stormy winds with the speed more than 18м/s are mostly detected from South-West.

Air's temperature regime. Average annual temperature – 14, 2°, the hottest month of the year –is August (average temperature – 23, 3°, maximal - 37, 3°). The coldest months are January, February (average temperature – 6-7°, minimal temperature -10°, in February).

Precipitation. Maximal amount of precipitation is in August-September (240-250 мм, on average, absolute minimum – 614 мм – in September). Absolute 24-hours maximum – 268 мм. Average annual amount of precipitation – 1661 мм.

Fogs. The fogs are detected in the spring in majority. The biggest average annual amount is detected in March – 3. There are 18 foggy days on average, in some years – up to 37 days.

3 HYDROLOGICAL CONDITIONS ANALYSIS

Tyagun. Tyagun possible in the heavy sea of 3-5 as well as in condition of the winds of the Western quarter with speed not less than 7-9м/. The most possible period is cold period of year – October-February. Only weak tyaguns and the first indicators are detected. Amplitude of the levels up to 10дм, and it is less in summer and more in the winter. The force of tyagun depends on the waves. In case the height of waves - 0,25м tyagun of 1 force is formed, the height of 1,25м – force 2 (3,5дм) and so on. The duration of tyaguns is about 12-83 hours. In case of heave sea force 1 the duration does not exceed 24 hours, in case of 4-5 – the duration between 66-72 hours.

Water Temperature. The highest temperature is detected in July-August (average temperature 24-25°, maximal - 29, 4°), the average temperature in the coldest months – January-March is 7-9°, the lowest – 2, 8° (February).

Salinity of water. Regime of water salinity plays an important role in hydrodynamical processes of the shore as well as in navigation in water area. Kulevi shore's average salinity of water is 14, 25

prm, which increases in the winter, at the lowest cost (the average salinity - more than 15 prm, maxsimum19, 7prm). The lowest salinity observed in May and June, during the spring flood Rion-discharge (average salinity - the order of 11-13 prm, the minimum 4,85 prm).

Heavy Sea. Wave situation in Port of Kulevi zone is determined by the wind waves and swell. The repeatedness of the West and North-West roughness are change only by 2, 5% and 9%. The Repeatedness of South-West in the spring-summer period increases up to 35-40%. In the cold period of the year on account of active influence of the west winds the wave regime increases. The height of the waves is about 2,0м which is 2, 0% in the winter period of the year. In the warm season, the maximum parameters of the waves are observed, mainly in the South-West of waves, the action of the South-Western, Southern and Western winds. The maximal parameters of the waves as a rule are connected with the wind-caused waves, which parameters are bigger than waves of swell. Average monthly repeat of the wind waves are 36-48%.

Currents. There are two major types of current: sea current, caused with the water circulation in the Black Sea (in the presented case – from South to North) and local shore current. In port Kulevi zone the currents are divide into wind, countervailing and draining. Countervailing currents are connected with the wind surges and directed to the sea. Draining currents are determined by the River Khobi current and directed to the sea. In case of permanent activity of heavy West and South-West (as a rule not less 12-18 hours) wave – along shore currents arise, which direction is constant, and speed reaches 1,0-1,5м/s. The maximum speed is in the region of the 5 meters isobaths. The wind with the constant speed direction and force is necessary for wind current formation (See Tab.10).

Table 10. Wind with the constant speed direction and force

Wind speed, м/s	5-10	10-15	15-20	More than 20
Time, hour		12-18	6-12	3-6	Less than 3

The analysis of the currents resulted in the following, period from 09.06.09:

1　When ship moves in the channel from receiving buoy to swinging room (swinging pool) zone №1 (Fig.4) she is influences the vector of the constant Northern current (sector $330^0 – 10^0$), force $0.6 – 1.3$ kn., depending on the season (See Tab.11).

Table 11. Wind direction depending on the season

January-March	350^0
April-June	350^0
July-September	10^0
October-December	330^0

2　In case the ship is moving by the stern to port, after swinging in swinging room (swinging pool) zone №2 (Fig.4) the vector of the South current stars its activity, which is formed in the result of the interrelation between the vectors of constant northern and western current of the river Khobi. Contributes to this in our opinion three factors:

– Encounter of tow, different waters streams – Northern Sea and Western fresh water;
– Deviation of the part of the vector of Western current to the South, in connection with formation of the local circulation with the general direction to the South, due to the artificial expansion of the river Khobi mouth and creation on the left bank of the fendering wall (vertical break from washing away and sliding sandy ground);
– Vector of the constant Northern current in interrelation with the wrecks which are in the Southern part of the channel form the local turn with the general direction to the South;

In other words the following takes place:
Vector of the Western current of the river Khobi striking the vector of the constant Northern current is partly taken by it and goes to the North, and its main part inclines from the bigger vector of the constant Northern current to the South.

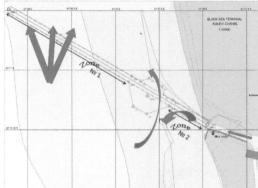

Figure 4. Influences the vector current when ship moves in the channel and swinging room

The dynamics of sediments. The dynamics of sediments in Port of Kulevi zone is determined by the firm drain of the river Khobi. The average of the firm drain is about 300 000 м3/year. But in the period of flood it may reach 450 кg/s (about 40 000 м3/24hours). The volume of sediment, brought from Rioni may reach 80-200 000 м3/year. The main collection of sediments, brought from the river Khobi is in the inner water area, as well as on the part of the channel 0-700м (bar formation). It must be noted that with the increase of the width of the water area from 300 to 500м the main part of sediment will accumulate on the inner water area and only tiny part of sediments will accumulate in the channel.

4 CONCLUSIONS

It is possible to conclude, that prospect of development of port Kulevi from the point of view of safety navigation, its economic feasibility and also effective operation of hydraulic engineering depend on the following:

- The main factor, which directly influences upon the level of safety of navigation in water area of Port of Kulevi is heavy sea.
- The major physical-geographic characteristics, which influence upon the wave processes, are force and direction of the wind – which determines the acceleration, time of the wind activity and the depth of the place.
- In area of port Kulevi dynamics of the sediments is caused, basically, by a firm drains of the river Khobi. The basic sedimentation of deposits occurs on internal water area, and also on a site entering channel in 0-700 meters, and it is necessary to carry out measurements of depths per 2 weeks and after each storm.
- The most dangerous, from the point of view of safety of navigation, are a wind of the western and southwest directions.
- The necessary level of safe navigation can be provided only by means of protective hydrotechnical constructions.
- It is necessary to conduct mathematical modelling of wave situations in different variants of protective measures for creation of protective hydro-

technical constructions. The theoretical and practical works have already been started and will be the topic of the following paper.

REFERENCES

Georgian Maritime Administration, The Regulations On establishing the traffic separation scheme, separation of sea corridors and maritime special areas in the territorial sea of Georgia. www.maradgeorgia.org

Dzhaoshvili S. "Hydrologic-morphological processes in mouth zone river Rioni and their anthropogenesis changes". Water resources, w.25, №2, 1998.

Varazashvili N. "Geological processes and the phenomena in a zone of construction of sea hydraulic engineering constructions and actions on improvement of a coastal situation". Engineering geology (excerpt). Academy of Sciences the USSR. Moscow, 1983.

Dzhaoshvili S. "New data about beach formed sediments of a coastal zone of Georgia". Water resources (excerpt). Academy of Sciences the USSR. Moscow, 1984.

Materials of supervision over elements of a hydrometeorological mode for the period 1971-1999. (Information of Poti mouth stations).

Bettes P. "Diffraction and refraction of surface waves using finite and infinite elements". Numerical Meth. London, 1997.

Bach H., Christiansen P. "Numerical investigations of creeping waves in water theory". ZAMM. New York, 2004.

Black Sea Terminal of Port Kulevi. http://www.kulevioilterminal.com

Gegenava A, Varshanidze N, Khaidarov G. " New Black Sea Terminal of Port Kulevi and it Navigating Features". TransNav 2009, Gdynia, 2009.

6. Hydro-meteorological Characteristics of the Montenegrin Coast

J. Ćurčić & S. Šoškić
Military Academy, Belgrade, Serbia

ABSTRACT: The paper describes the mathematical, physical, geographical and hydrological-meteorological characteristics of the Montenegrin coast, as well as their impact on the sailing boat.
This paper is part of a study wich was conducted from July 2000 to august 2006 with officers on the ships Navy Serbia and Montenegro.

1 COAST OF MONTENEGRO

1.1 *Type area*

Montenegrin coast is a complex entity of three ambient spaces: coastal sea (internal waters and territorial sea), coastal belt and the airspace above them. For this paper need and influence of meteorological conditions, include the land in the hinterland of the coastal edge.

Coast of Montenegro has an irregular semiconstrict shape. Extends between 41°50'30"N and 42°34'48"N and 18°25'30"E and 19°24'42"E, between Croatia and Bosnia and Herzegovina in the northwest and Albania in the southeast. Width of the Montenegrin coast is about 90-95km, and the depth of it is 10 to 50km. It has a peripheral position in relation to waterways in the Adriatic Sea that are closer to the coast of Italy.

Total area is 3529.9km2 and consists of the coastal sea area of 2,393.9km2 (2,047.2km2 of territorial sea and internal waters area of 346.7km2), coastal zone area of 1.136km2 and is part of Skadar Lake of Montenegrin area of 229.8km2.

1.2 *The relief of the Montenegrin coast*

Montenegrin coast stretches between the Adriatic coast and mountainous areas of relief. It is highly karst landscape as it occurs rarely in the world, characterized by the fact that immediately follows the coast, at the greater or lesser distance, steeply rising young wreath mountains: Orijen (1893m), Lovćen (Štirovnik-1748m) and Rumija (1594m) that stretch parallel to the coastline.

Figure 1. Satellite image of the Gulf of Kotor

1.3 *Characteristics of the Montenegrin coast*

The length of coast line is 293.6km, and the Gulf of Kotor is 105km. The length of air line from the mouth of Bojana River to the Cape Oštro is 96km. The coefficient of indentation of the coast is 3.06. The coast is open; and a small number of islands, rocks and reefs, located off the coast, is of little importance in the protection of the coastal zone.

Older limestones that make up this section are separated from the flysch zone by rift. In Boka occurs another flysch zone, and that is the Inner zone Morinj-Kotor. With this zone, to the northeast, rises the rift part of Mesozoic sediments.

The composition and structure of only one part of the coast between Budva and Bar is very complex and composed of Eocene flysch and Mesozoic limestone reefs. The direction of the layers is Dinaric. Also, south of Petrovac sinks under the sea the other flysch zone and calcareous ridge southeast of it. The coast is still composed of Triar limestone.

From Bar to Ulcinj again occurs flysch zone which is divided by a ridge of limestone composed of Eocene and Cretaceous limestone. From Ulcinj to the mouth of Bojana coastal zone is under the influence of the delta of Bojana, sandy, with very little gravel.

Figure 2. Displaying the Bay and the Gulf Trašte

On the Montenegrin coast largest and most important gulfs are: Boka, Trašte, Budva, Bar and cove Valdanos. Gulf of Kotor consist of four secondary bays (Kotor, Risan, Tivat and Herceg Novi) linked with Strait (Strait Kumbor and Strait Verige).

Significant part of coast line, with total length of 293.6km, belongs to beaches, of 17km are rocky beach and 56km are sandy beaches. In Gulf of Kotor beaches are mostly small, stone, concrete or sand based in some places. Outside of the Gulf, i.e. southern part of the Peninsula Lustica Cape Mirište until Pastrovic (Budva) has plenty of beaches. Further to the southeast along the Montenegrin coast major beaches and bays are: Slovenska beach in length of 1.300m, Bečići (1.800m), Buljarica (1.800m), Bay of Čanj and beach, Sutomore beach (1.300m), Šušanj (1.000m), cove Valdanos and Ulcinj sandy beaches in total length of 24km of which Ulcinj's Velika plaza of 13.5 km is the longest, and it can accommodate up to 100,000 people, and the sea bottom in front of it is sandy.

2 HYDROLOGICAL AND METEOROLOGICAL CHARACTERISTICS OF THE MONTENEGRIN COAST

Hydrological and meteorological characteristics must be considered when planning navigation through Adriatic coast in Montenegro: the relief of the seabed, sea temperature, salinity, density, transparency, currents, tides, the direction and strength of wind, sea state, air temperature, air pressure, cloud and fog and rainfall.

2.1 The bottom is relief of the southern Adriatic

Basin of the Adriatic Sea is a young foundation, from the Early Tertiary. The Adriatic Sea obtained present form not until the end of Diluvium.

Longitudinal relief of the Adriatic Sea is quite diverse. The shallowest northern part is the part where the depth is less than 50m. South of this line depth is slightly increasing, but still less than 100m. Obstacle Palagruža south of the sea floor lowers and forms South Jabuka valley with a maximum depth of 1.233m. Depths towards the Strait of Otranto again decrease and in the Otranto make 741m.

Figure 3 shows the vertical profile of depth in the southern Adriatic on connector port Ulcinj with depth of 1.233m ($\varphi = 41°41'N$, $\lambda = 18°16'E$). For the slice was used nautical chart 300-35. It is evident that the depth slowly increased to the extent of the continental shelf (200m) on distance from the coast about 50 km, and the later rapid growth to a depth of 1.233m.

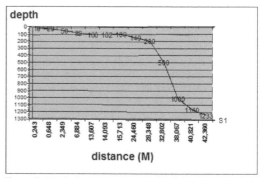

Figure 3. Vertical cross-section depth of the junction of the southern Adriatic port of Ulcinj - 1233m depth

Because of these characteristics of the bottom relief in the territorial sea is no danger to navigation on fairway.

2.2 Sea Temperature

The highest and the lowest sea temperature is delayed on average, one month after extreme values of air temperature; in February temperature is the lowest and the highest in August.

In the southern Adriatic prevailing surface temperature is 12°-13°C. Along the eastern coast the average temperature is 10°-11°C, and along the west one degree lower.

For much of the northern Adriatic Sea temperature in August is 22°C, in middle Adriatic Sea average temperature is 24°C and in the southern it is 25°C.

Figure 4. Display isotherms of the Adriatic Sea

Along the eastern coast annual fluctuation of the sea surface temperature is 14°-18°C, on the open sea average fluctuation is 12°C. Daily fluctuations in surface temperatures are highest in summer and amount to 1.5°C. In February, as well as during the winter, mainly in the Adriatic there is isotherm. The exception is the Jabuka Pit where the bottom temperature is around 11°C, for about 2°C lower on the surface. Daily temperature changes in the Adriatic are felt to a depth of 30m, annual to 300m.

Changes in sea temperature do not have significant influence on the propagation of sound through seawater, as well as errors in determining the depth of with the sonar.

2.3 Salinity

The salinity of the open Adriatic has an average of 38.5‰, it is higher in southern (38.7‰) than in the central Adriatic Sea (38.3‰) and is sustained throughout the year. In middle and southern part there are two different peaks of which the September peak is primary and of February - secondary, two minimum of which the May is primary and secondary is in December.

2.4 The density of sea water

Water mass is not homogeneous from surface to the bottom. Density depends primarily on salinity, temperature and pressure. Density must be expressed by specific weight. Increasing salinity increases density too, because of bigger presence of salt.

Table 1: density dependence on the salinity

salinity (‰)	30	32	34	36	38	40
density (kg)	1,024	1,025	1,027	1,028	1,030	1,032

Source: Pajković M., Oceanography, pp. 41

In the Adriatic the flow isopikni (line of the same density) approximately coinciding with the isohaline (line of the same salinity). To salinity and density, significantly affects the inflow of sea water and underwater springs.

2.5 Transparency of the sea

Transparency has the function of the intensity of illumination of the sea surface, the physical properties of seawater, sea state and observer's visual abilities. Transparency is weak in the area of mixing water masses of different physical properties it is. Reduced by the amount of material that rivers bring in during heavy rainfalls. State of the sea surface significantly affects the transparency so that sea with index 2 by Beaufort can be completely opaque.

The Adriatic Sea has a lower transparency than Ionian and Mediterranean sea as a whole. The maximum values of transparency were recorded in the South Adriatic Basin, and are up to 56m.

Shimmer and color are closely related with transparency as a hydrological feature of the sea. Color of the sea near major port centers is dirty, gray, and at mouth of rivers into the sea is yellow. Color depends on the brightness of the sky and the color of the sea floor. In shallow waters blue turns into green and light green, and warns sailors of caution in navigating among other navigational aids, if they are defective or if they are .

2.6 Ocean currents

In the Adriatic Sea we can distinguish four types of currents: general, windy, tidal and drift. General current entering through the Otranto and moving to the north along the coast of Albania and Montenegro. The average speed of currents along the eastern coast in the south is about one knot. Tidal stream are changing every 6 hours, and the direction may be agreed with the general or against this flow. Drift currents are local and general character of the Adriatic, which depends on the period of oscillation, and shapes the bay in which drift arise.

Depth of flow was affected by the configuration of the bottom and winter cooling water masses in the northern Adriatic. Aggravated by winter water flows at the bottom above the Threshold of Otranto north-Adriatic winter water goes into the Ionian Sea filling par holly east-Mediterranean Basin. As compensation, there is increased flow to the area from the Ionian into the Adriatic Sea with a tendency to compensate denivelation in northern Adriatic.

In the spring and especially in the summer the situation is completely changed. Due to snow melt and increased inflow of water, level in the northern Adriatic Sea is increasing; the outflow is carried out down the Apennine side.

2.7 Tides

Tides in the Adriatic Sea are in dependent phenomenon. Tidal wave enters the Adriatic from the Ionian Sea every 12.4 hours (half lunar day). Adriatic sea has a half-day type of sea change, i.e. two high and two low water during the day.

Table 2: Summary of sea change in 2001

January		February		March	
MHW:	121,6	MHW:	110,5	MHW:	122,4
MLW:	99,5	MLW:	88,0	MLW:	97,9
Amplitude:	22,1	Amplitude:	22,5	Amplitude:	24,6
April		May		June	
MHW:	108,9	MHW:	110,0	MHW:	103,9
MLW:	85,6	MLW:	87,1	MLW:	83,2
Amplitude:	23,3	Amplitude:	22,8	Amplitude:	20,7
July		August		September	
MHW:	105,1	MHW:	106,5	MHW:	117,7
MLW:	83,7	MLW:	84,5	MLW:	94,3
Amplitude:	21,4	Amplitude:	22,0	Amplitude:	23,4
Octobar		November		December	
MHW:	111,1	MHW:	112,3	MHW:	105,0
MLW:	86,8	MLW:	88,8	MLW:	83,6
Amplitude:	24,3	Amplitude:	23,5	Amplitude:	21,4

Source: Report on mareographic the observations Bar for 2001, HI of Navy, 2002.

Table No 2 shows that the maximum amplitude measured in 2001 were 39.2 cm. This measurement show that amplitude has no effect on the accuracy of position and anchoring the ship.

2.8 Wind direction and strength

The most frequent winds are the bora, jugo and the mistral.

Bora is a dry and cold wind, blowing from the northeast in general direction. Bora is characterized by low relative humidity and air temperature, with good visibility and mostly clear skies.

As a sign of the phenomenon, i.e. beginning of the storm, there is a "cap" of clouds on mountain tops and passes (even when short-term planning should be given to local signs of the coming time).

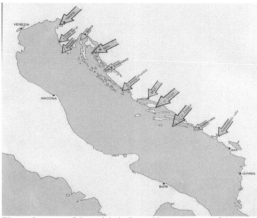

Figure 5. Areas of the Adriatic Sea with a very powerful storm strikes

Jugo is a warm and moist wind which blows from the southeast. It occurs throughout the year, but is more common from fall to late winter. Jugo is a strong and powerful wind that creates high waves and usually brings rain. Although it is a strong wind, it is less dangerous than bora because it needs some time to reach its maximum (36 to 48 hours). It reaches gale force after the third day of continuous blowing.

Mistral is a wind from the northwest direction, which is characteristic for the warmer part of the year it. Will rarely occur in the winter, but if it stars blowing it is sign of extremely bad weather.

Jugo and mistral are characteristic for Herceg Novi. From the above one can not say with certainty can will be expected during the year in a statistical analysis. Precisely because of this, Table No 3 be of a big help because it shows the number of days when the wind is stronger than 6 Beaufort (12.3m/s).

Table 3: Duration of the wind to 6 Beaufort

6 Bf	max	min	aver.
jan	15.0	0.4	4.0
feb	15.0	0.6	4.3
mar	17.5	0.7	5.3
apr	16.2	0.3	5.1
may	13.8	0.1	3.8
jun	13.9	0.1	3.7
jul	11.5	0.1	3.4
aug	11.6	0.1	2.9
sep	12.9	0.2	2.5
oct	15.2	0.3	3.3
nov	16.4	0.3	4.0
dec	13.9	0.4	4.0
year	172.9	6.2	46.2

Source: Climatologically data RHMI of Montenegro with stations at sea, Podgorica, 2005.

2.9 Sea condition

Sea condition is a complex concept. The basic characteristics that show the state of the sea are direction, wave height and length. Although the wave height and length are measurable size, the state of the sea has the empirical size and estimated empirically ("roughly"). In Beaufort scale is a description of the evaluation of the sea.

The coast of the Gulf of Kotor to the mouth of the River Bojana is exposed to winds from the second and third quadrants, which are blowing strong and make rough sea, even when the wind is moderate. A northwest wind is very cold and makes big waves. Jugo is, due to the direction of extention of the shore, forced to change direction to the southeast, and therefore the waves are moving in this direction. The highest waves occur from the northeast, southeast and south.

In front of the Port of Bar southwest and west winds create rough sea. Port of Ulcinj is fully exposed to south winds and waves.

Figure 6. Description of the state has strength 2 in Beaufort

The averagely highest waves comes from south. In most part of year its strength can be 3 to 5Bf more than, three days during a month.

From the southeast the waves usually occur in April (seven days, the strength of 3 to 6Bf), November (nine days, the strength of 4 to 5Bf), and in December (six days on volume 4 to 6Bf).

Waves from the northwest direction occur in the summer (from May to August), on average, eight days in a month, volume 3 to 5Bf, with the exception of July when it may occur as many as eleven days.

From the northeast in the period from September to March, in the southeastern part of the sea waves occur at strength of 3 to 6Bf, four days a month. In the period January-February, the sea has strenght of volume 4 to 6Bf, six days a month.

2.10 Air temperature

On the Montenegrin coast is mainly represented by the Adriatic-Mediterranean climate with short mild winters and long warm summers. In the area of Boka mean January temperature is 6.9°C to 8.4°C, mean July temperature is 23.4°C to 24.3°C, and the mean annual temperature is 15.5°C to 15.8°C.

Table 5: Average monthly air temperature

	H. Novi	Budva	Bar	Ulcinj
jan	8.2	8.4	8.3	6.9
feb	8.8	8.9	9.0	8.0
mar	10.6	10.7	10.5	10.3
apr	13.6	13.7	13.4	13.6
may	17.8	17.8	17.6	17.8
jun	21.7	21.7	21.3	21.8
jul	24.3	24.0	23.4	24.3
aug	24.1	23.6	23.1	24.3
sep	20.8	20.6	20.3	21.0
oct	16.6	16.7	16.6	16.7
nov	12.8	13.3	13.2	12.4
dec	9.9	10.2	10.1	8.9
aver.	15.8	15.8	15.6	15.5

Source: Climatologically data RHMI of Montenegro with stations at sea, Podgorica, 2005.

High humidity and high temperature can lead to fatigue of the crew and a temporary inability to work on the deck.

2.11 Air pressure

Air pressure is taken as the force with which the air acts on the surface, under force of gravity. Normal atmospheric pressure at sea level is 1.013mbar at 0° C and 45°N latitude.

Table 6: Mean monthly pressure in hPa

	H. Novi	Budva	Bar	Ulcinj
jan	1015.1	1016.1	1015.7	1013.1
feb	1013.5	1014.3	1014.0	1011.7
mar	1013.8	1014.4	1014.1	1011.6
apr	1012.0	1012.6	1012.2	1010.2
may	1013.5	1014.0	1013.6	1011.2
jun	1013.6	1014.0	1013.7	1011.4
jul	1012.8	1013.5	1013.0	1010.5
aug	1013.0	1013.6	1013.1	1010.6
sep	1015.6	1016.2	1015.7	1013.3
oct	1016.6	1017.0	1016.5	1014.3
nov	1016.0	1016.6	1016.1	1014.0
dec	1014.7	1015.5	1015.0	1013.2
aver.	1014.2	1014.8	1014.4	1012.1

Source: Climatologically data RHMI of Montenegro with stations at sea, Podgorica, 2005.

From table No. 6 can be seen that the mean pressures in the southern Adriatic ports is around the values of normal atmospheric pressure. This leads to the conclusion that the source region of cyclones and anticyclones far away from the waters of the Montenegrin coast and made an indirect impact on the time, so you must take into account the wider geographical area from the Adriatic Sea.

2.12 Cloud and fog

Cloudiness is part of the visible sky covered with clouds and it is estimated visually roughly. In meteorology it is measured in tens. In the Air Force in eights, and in Shipping in the quarters.

Cloudiness on the Adriatic is generally small. The annual average in the southern part is 4.6 to 5.1 tenths. The biggest cloudiness is in December and January. In the summer it is much smaller. In particular, a small cloudiness in July and August (2.4 to 2.5 tenths).

Table 7: Mean monthly cloudiness in tenths (1/10)

	H. Novi	Budva	Bar	Ulcinj
jan	6.2	6.2	6.1	5.9
feb	6.4	6.3	6.2	6.1
mar	6.0	5.8	5.8	5.7
apr	5.8	5.6	5.5	5.4
may	5.2	5.0	4.9	4.7
jun	4.1	3.7	3.7	3.5
jul	2.6	2.4	2.3	2.1
aug	3.2	2.4	2.3	2.1
sep	3.7	3.5	3.4	3.2
oct	4.7	4.7	4.5	4.4
nov	6.5	6.5	6.3	6.1
dec	6.4	6.3	6.1	5.9
aver.	5.1	4.9	4.8	4.6

Source: Climatologically data RHMI of Montenegro with stations at sea, Podgorica, 2005.

The largest cloudiness is from November to February when it is 6.1 to 6.4 tenths. Maximum average monthly cloudiness is (6.5) in November, above Budva and Herceg Novi, while it the lowest is (2.1) in July and August over the Ulcinj.

Gloomy days are more frequent in the winter. From October to March, averagely there are 11 to 13 of gloomy days. The are most groomy days in Budva and least in Ulcinj. There are most clear days in July and August (over 17).

Table 8: Number of days with fog

	H. Novi	Budva	Bar	Ulcinj
jan	0.0	0.0	0.1	0.1
feb	0.1	0.0	0.0	0.1
mar	0.3	0.4	0.2	0.9
apr	0.3	0.2	0.3	0.6
may	0.3	0.3	0.5	1.3
jun	0.0	0.1	0.2	0.8
jul	0.1	0.2	0.1	0.9
aug	0.0	0.1	0.4	1.3
sep	0.1	0.5	0.2	1.7
oct	0.1	0.0	0.0	0.1
nov	0.2	0.1	0.1	0.1
dec	0.0	0.0	0.0	0.1
sum	1.5	1.9	2.1	8.0

Source: Climatologically data RHMI of Montenegro with stations at sea, Podgorica, 2005.

Fog occurs on average 3.4 days per year, maximum 8 days in Ulcinj. In Herceg Novi it occurs only 1.5 days per year. It usually occurs from March to April and in August and September (about half of the month).

2.13 The amount of rainfall

The coastal relief and circulation of air masses affect the distribution of rainfall. High Dinaric Alps stop air masses brought by winds from the south and create rain clouds.

Rainiest region in Europe is in the field Krivošija where annual rainfall is around 5000mm of rainfall. The frequency of rainfall is every second or third day. Rainy days are seen in November (15 days), and then from December to April (10 to 15 days a month). In May, September and October, rainy days there are 5 to 10. The smallest number of rainy days is in July and August, less than 5. Ice is almost nonexistent. On average there are 3.7 days per year.

The amount of rainfall is proportional to the number of rainy days (Table No 9).

Table 9: Mean monthly rainfall

	H. Novi	Budva	Bar	Ulcinj
jan	256.3	173.2	171.5	174.3
feb	213.7	161.1	155.0	145.8
mar	196.8	146.1	131.8	120.1
apr	164.0	124.5	124.0	110.4
may	114.2	100.6	102.1	74.0
jun	63.9	60.7	59.6	54.2
jul	46.2	43.9	38.4	28.8
aug	83.7	61.9	52.1	45.9
sep	168.1	126.7	117.5	96.6
oct	184.0	167.7	136.4	139.4
nov	289.0	210.9	203.7	182.1
dec	246.9	181.6	176.1	163.9
sum	2.026.8	1.558.9	1.468.3	1.335.6

Source: Climatologically data RHMI of Montenegro with stations at sea, Podgorica, 2005.

CONCLUSION

The paper presents briefly describes the hydrological and meteorological characteristics of the Montenegrin coast. In fact, the hydrological and meteorological characteristics conducive to sailing and navigation on the Montenegrin coast.

On safety of navigation does not affect everyone equally. The biggest influence is that wind and sea condition. They can impede or completely stop sailing.

Pressure and temperature have an indirect impact on the crew of the ship. tides do not have a significant impact on navigation or anchoring. maximum amplitude recorded at a tide gauge in the bar is approximately 1 m.

REFERENCES

Climatologically data RHMI of Montenegro with stations at sea, Podgorica, 2005.

Ćurčić J., The basis of marine meteorology, Military academy, Belgrade, 2000.

Ćurčić J.: Hydrometeorological security in the preparation of Navy operations, Master's thesis, Military academy, Belgrade, 2008.

Marjanović R., General military geography, Military academy, Belgrade,

Matas M., Mediterranean, Schoolbook, Zagreb, 1981.

Mišović S., The military estimates the Yugoslav battlefields, Ph.D. thesis, Military academy, Belgrade, 1996.

Pajković M., Oceanography, Scientific book, Belgrade, 1954.

Report on mareographic the observations Bar for 2001, HI of Navy, Lepetane, 2002.

Rodić D., Geography of Yugoslavia I, Scientific book, Belgrade, 1987.

Tešić M., The Naval military geography, DSNO, Belgrade, 1968.

WMO, Polar orbiting satellites and applications to marine meteorology and oceanography, 1996.

Ice Navigation

7. Ship's Navigational Safety in the Arctic Unsurveyed Regions

T. Pastusiak

Gdynia Maritime University, Gdynia, Poland

ABSTRACT: High traffic of the vessels in many regions of the world pressed maritime nations to issue good quality nautical charts. Vessels could proceed safely on planned voyage using nautical chart and GPS position receiver. Above popular assumptions were right in well recognized and charted regions. But some regions were not sufficiently surveyed or not surveyed at all. In this case position fixing system was useless. The only way was to follow the vessel's hydroacoustic equipment to find out safe route in between dangers.

The goal of the author was to settle matters of the unsurveyed regions. First question was quality of the information on charts and role of the vessel's autonomous hydroacoustic equipment in safety of the navigation. Second question were safety parameters kept by the research vessel.

1 INTRODUCTION

The popular assumption is that a safety of the vessel is fulfilled by having in vessel's disposal a navigational chart in proper scale and a device or a method fixing position on this chart. By this way is possible to establish the main criterion of the navigational safety in regions well recognized - distance to the dangerous objects. In this case the object is univocally defined and charted.

The goal of the author is to establish methods for assessment of the vessel's safety during planning and monitoring of voyage in unsurveyed or poorly surveyed regions in clear and simple manner that fulfill requirements of governing regulations (IMO 1993). It includes the navigational information in world resources of charts and the autonomous equipment possible to be on board the vessel. One identifies relations corresponding to the navigational safety based on field data.

2 MEANS TO SUPPORT THE NAVIGATION

The navigational support of the voyage was divided into internal and external navigational information. Internal one was related to the ship's own technical devices. External one was related to the charts, pilot books or other information.

2.1 *External methods to support the navigation*

The main sources of the information for the safe navigation were sea charts contents. Rest of the information came from the various nautical publications. Usefulness of the charts was assessed at first approach by theirs scale and reliability of the content (Pastusiak 2010).

2.1.1 *Scale of charts*

The application of the charts for navigational purposes was closely correlated with theirs scale. The electronic chart catalogues (Jeppesen Norway A/S 2010; Primar Stavanger 2010; Transas Marine Ltd 2010; UKHO 2010) and the internet chart catalogues (IC-ENC, http://www.ic-enc.org, 23-Mar-2011; NOAA, http://charts.noaa.gov, 16-Jan-10; NOAA, http://www.nauticalcharts.noaa.gov, 16-Jan-10; IC-ENC, http://ic-enc.org; Garmin Ltd, http://www8.garmin.com, 08-Jan-10; Jeppesen Norway A/S, http://www.c-map.no, 25-Jan-10; ChartWorld GmbH, http://www.chartworld.com, 29-Jan-10) introduced division mostly in 6 groups of charts. It was related to kind of voyage, details of the information included and the scale of chart (Weintrit 2009). The scale of chart was correlated with position error of features placed on chart. It was 0.3 millimetres in lineal measure. In Table 1 there are presented the position errors of the placed information (features) related to the worst scales of charts in the group.

Table 1. Groups of charts and position error of charted features.

Group (band)	Scale	Position error of charted feature (m)
Overview	1:700,000 or smaller	700 or more
General	1;180,000 to 1:350,000	105
Coastal	1:75,000 to 1:180,000	54
Approach	1:12,500 to 1:45,000	13.5
Harbour	1:8,000 to 1:22,000	6.6
Berthing	1:4,000 or greater	1.2

2.1.2 Reliability of chart content

Reliability of charts content was described by date of a survey when source data came from. Actually used descriptions like "unsurveyed" region, "poorly examined", "inaccurately examined", "fully examined" should be correlated with presently being introduced meaning like Zones of Confidence ZOC (Gale 2009; UKHO 2004). Zones of Confidence referred to detection and quality of the measurement of the features on a seabed. Important matter was probability of missing (not placing) a navigational danger on a chart . Zones of Confidence were not implemented on all charts till now. On many electronic charts of not well surveyed regions placed ZOC category "U" that means "unclassified". The vessels should use best scale charts for the intended voyage. The world charts resources were searched in relation to Murchisonfjorden region at Nordaustlandet.

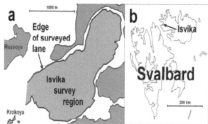

Figure 1. (a) Surveyed region in Isvika (contour line indicates edge of surveyed lane). (b) Isvika survey region on Svalbard

Table 2. Coverage of Isvika region by charts.

Source	Scale / Bands	SOLAS	Kind of chart
UKHO	general 1:600,000	Official	paper, ARCS
Norwegian HO	1:100,000	Official	Paper
Russia GUNiO	1:200,000	Official	Paper
AVCS	Transit	Official	Electronic
ECDIS Service	Full =1:600.000	Official	Electronic
Primar	no coverage	-----------	------------------
Transas Marine	1:200,000	Unofficial	Electronic TX-97
Garmin Bluecharts	1:100,000	Unofficial	Electronic products Garmin
Jeppesen Marine C-MAP	1:100,000	Unofficial	Electronic products NT, MAX, MAX PRO
Jeppesen Marine C-MAP	1:1,500,000	Official	Electronic CM-93/3
SevenCs GmbH	Harbour	Unofficial	Electronic Navionics ENC

2.1.3 Sources of origin of the chart

For purpose of this work reviewed, taken into consideration and subsequently divided charts as follows: official, unofficial, „other – bathymetric" and „other - non bathymetric". Official charts fulfilled requirements of SOLAS, Chapter V, Regulation 2.2 (IMO 2004) that states "Nautical chart or Nautical publication is a special purpose map or book, or a specially compiled database from which such a map or book is derived that is issued officially by or on the authority of a Government, authorised Hydrographic Office or other relevant government institution and is designed to meet the requirements of marine navigation". Official charts published by Hydrographic Offices guaranted systematic updates of the informational content according to IMO requirements. Zones of Confidence scale should be available on the chart.

Unofficial charts were of commercial destination. Theirs informational content had same source of origin like charts issued by Hydrographic Offices. Unofficial charts not fulfilled SOLAS requirements and not guaranted systematic updates of the informational content. These charts frequently contain additional commercial information. Vessels operating on unofficial charts are to have and use also up-to-dated official charts - at least paper ones.

„Other – bathymetric" unofficial charts were of scientific value. The goal of the authors was the most reliable presentation of depths and sea bottom relief. Source materials were made frequently without taking into consideration standards of related to hydrographic surveys described in IHO publication (IHO 2008) by the persons not being qualified in the hydrography discipline nor production of official sea charts. Such charts not included in most cases corrections for sea level in relation to Chart Datum nor corrections for vertical location of sounder or echosounder transducer. Accuracy of sounding was not estimated nor included in depth reduction. However „other – bathymetric" charts were related to hydrographic niches and sometimes were valuable source of information about sea bottom relief in the region of interest. Due to lack of better sources of hydrographic information these charts could be usefull for the initial voyage planning of hydro-graphic surveys. Informational content allowed to grant them class from „Coastal" till „Approach". Appointment of ZOC class for each „other - bathymetric" unofficial chart required individual assessment.

„Other – non bathymetric" charts were of scientific value. Theirs authors not planned reliable presentation of depths nor sea bottom relief. Sources of information related to the sea bottom were in most cases unknown. However these charts contained informations that allowed to give them ZOC class „Overview". For the voyage planning purposes ZOC scale on „other – non bathymetric" charts was not so important.

Reliability of the information content was attributed to the new scale ZOC that replaced informations about date of last hydrographic survey in the mentio-ned region. Assumed, that implementation of new scale of reliability of informational content on the charts requires prolonged period of time. It was due to necessity to re-assess date of hydrographic survey and corellated informations on actual charts that not corresponded with new precise scale of ZOC.

During process of voyage planning in the unsurveyed or poorly surveyed regions should be taken into consideration the coverage of the region of interest by charts for navigational purposes take into account all three informative elements: the scale of a chart, the scale of reliability ZOC and the reliability of the sources of origin of the chart information.

External methods to support the navigation can be estimated by reviewing available world charts resources and the charts possessed by the vessel. Introduced quality scale of support (Table 3) to evaluate external support to the navigation on board the vessel. The norm was the scale of chart comparable to the planned kind of the navigation (UKHO 2009; UKHO 2010; Jeppesen Norway A/S 2010; IC-ENC 2010; ChartWorld GmbH 2010; Primar Stavanger 2009; Weintrit 2009).

Table 3. Assessment of external support.

Kind of charts	Scale same with norm	One level lower then norm	Two levels lower then norm	Three levels lower then norm	Four levels lower then norm	Five levels of lower chart then norm	Lack
Official charts	6	5	4	3	2	1	0
Unofficial charts	6	5	4	3	2	1	0
"Other charts – bathymetric"	6	5	4	3	2	1	0
"Other charts – non-bathymetric"	6	5	4	3	2	1	0
Sum of scores	Maximum possible 24 scores						

For easy assessment of the external support to the navigation available on board a vessel introduced relative coefficient of the external support Ce expressed by Equation 1:

$$C_e = \frac{(O + U + B + N) \cdot 100}{24} \qquad (1)$$

where C_e - the coefficient of the external protection (%); O - quality rating support by the official charts in scores from 0 to 6; U - quality rating support by the unofficial charts in scores from 0 to 6; B - quality rating support by the "other charts - bathymetric, in scores from 0 to 6; N - quality rating support by the "other non-bathymetric charts" in scores from 0 to 6.

The comparison in between potential and actual support on board the vessel can indicate possibility and/or necessity of improvement of the external support quality.

The survey region of Isvika was situated in the South Eastern part of Murchisonfjorden located on Nordaustlandet (79°58'N, 18°33'E). The bottom of Isvika region was rocky, partly coated by a layer of sediments of glacial origin. From the external sources of the information (UKHO 2007; The Norwegian Hydrographic Service and Norwegian Polar Research Institute 1990) found that the surrounding region not passed any systematical survey. The ships should navigate with considerable caution because the sea bottom is very irregular. To be taken into consideration the existence of not detected dangerous banks. Ascertained existence of almost vertical changes of depth. Even at depths 50 -100 meters can appear small depths in vicinity. It requires special caution. The distances to the visible apparent danger (coast line) on the radar screen during surveys were about 0.05 nautical miles (Fig. 2a). However, the coast line was not the closest dangerous feature. The closest dangers were unknown small depths in close vicinity of the vessel (Fig. 2b). Reviewing the above mentioned external information ascertained the proper scale of a chart required for survey works in Isvika region as 1:10,000. It corresponded to the group "Harbour".

The official paper chart (Statens Kartverk 2001) shown reliable isobaths, features and coast line. The scale of the chart and the informational content not assured 100% of navigational safety. The official electronic chart of Transas Marine in scale 1:12,500,000 was not qualified to support navigation. The unofficial electronic chart Garmin Bluechart shown isobaths, features and coast line properly, but the scale of the chart and the informational content not assured 100% of navigational safety.

Insufficient external information on the charts required to support the navigation in poorly surveyed region (an partly not surveyed at all) of Murchisonfjorden including Kinnvika and Isvika (The Norwegian Hydrographic Service and Norwegian Polar Research Institute 1990) with the autonomous ship's internal methods detecting dangers to the navigation.

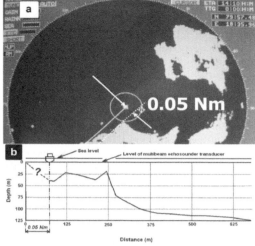

Figure 2. (a) Distance to danger on radar image. (b) Transverse depths profile at radar position

2.2 Internal methods to support navigation

Internal methods to support the navigation of the vessel were based on possessed by the vessel technical resources. They allowed autonomous detection of underwater dangers. Advantages and disadvantages of each method not clarified superiority any of below mentioned methods.

2.2.1 Sonar looking forward

The sonar looking forward allowed detection of underwater objects (features) in front of the vessel. The image of the situation was presented on heading in vertical and horizontal sections. In case sonar looking forward was only one electro acoustic device being on board vessel, one could continue a safe voyage in any direction.

2.2.2 Multibeam echosounder

The multibeam echosounder detected underwater objects (features) in transverse plane of the vessel. It not informed about the situation in front of the vessel. In case the multibeam echosounder was only one electro acoustic device being on board the vessel in unsurveyed regions, one could continue safe voyage across planned direction of the voyage. It required proceed along the lanes of previous measurements of the multibeam echosounder.

2.2.3 Single-beam echosounder

The single-beam echosounder detected underwater objects along perpendicular line under the vessel. It not informed about the situation in front of the vessel. In case the single-beam echosounder was only one electro acoustic device being on board the vessel in unsurveyed regions, one could extrapolate a distance to the potential underwater danger from

tendency of depth changes. The single-beam echosounder was not a fully autonomous device nor assured 100% safety of the navigation.

2.2.4 Echosounder on boat proceeding in front of the vessel

The boat proceeding in front of the vessel was equipped with single-beam echosounder. Results of this method were very similar to sonar looking forward. Safety output depended on qualifications of the boat crew and cooperation in between the boat and the vessel. It required good radio information exchange.

Quality of internal methods to support the navigation ascertained by reviewing equipment possessed by the vessel. Proposed scale was presented in Table 4.

Table 4 . Assessment of internal support.

Device being aboard	Efficient and reliable	Efficient and not reliable	Inefficient or lack device
Sonar looking forward	2	1	0
Multibeam echosounder	2	1	0
Single-beam echosounder	2	1	0
Echosounder on boat proceeding forward	2	1	0
Sum of scores	Maximum possible 8 scores		

For easy assessment of internal support to the navigation available on board the vessel was introduced relative coefficient of internal support C_i expressed by Equation 2:

$$C_i = \frac{(F + M + S + R) \cdot 100}{8} \qquad (2)$$

where C_i – the coefficient of the internal support (%) ; F - quality rating support by the sonar looking forward in scores from 0 to 2; M - estimation of support by the multibeam echosounder in scores from 0 to 2; S - estimation of support by the single-beam echosounder in scores from 0 to 2; R – estimation of support by the boat with the echosounder in scores from 0 to 2.

3 ESTIMATION OF THE DISTANCE TO THE DANGERS WITH THE MULTIBEAM ECHOSOUNDER

The assessment of navigational safety in the regions well recognized and charted was made in relation to the superficial and underwater dangers plotted on the sea charts. The information about the dangers on Svalbard not existed or was not sufficient or was not reliable. The assumed "danger" was unknown region ("blank place") out of the edge of not processed information from the multibeam echosounder (Fig.3a). The criterion of danger was the distance to this edge (Fig.3b). The navigator made continuous interpretation of multibeam echosounder image. Curvature of sea bottom was presented by serial of dots. The contiguous line to the most external dots allowed extrapolation of the sea bottom curvature up to sea surface. It gave additional reserve to the expected danger. In some cases was not easy to identify fallacious dots from whole sea bottom line. In dependence from the navigator decision various contiguous lines could be taken into consideration. This led to receive various reserve of expected distance to the "danger".

The sea charts of Murchisonfjorden region not assured safe navigation. These charts based on the information from the paper chart in scale 1:100.000 (Statens Kartverk 2001). Theirs information content not shown all features discovered by the multibeam echosounder. In some cases the sea charts shown inadequate locations of the coast line and the bottom features. The British and Norwegian pilot publications contained very limited information. Same time these publications advised mariners about almost vertical high changes of depths in the western part of Murchisonfjorden (The Norwegian Hydrographic Service and Norwegian Polar Research Institute 1990). Taking into considerations the International Hydrographic Organization (IHO) regulations, soundings and changes of depths described above, the regions of Isvika and Murchisonfjorden should be treated as insufficiently surveyed and partly as unsurveyed regions (IHO 1994; IHO 2009; The Norwegian Hydrographic Service and Norwegian Polar Research Institute 1990).

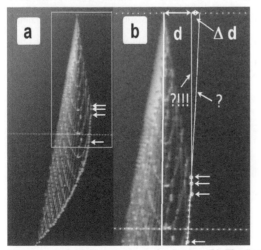

Figure 3. (a) Multibeam echosounder image, (b) The "unknown" area on multibeam echosounder image and probable distance to danger"

3.1 Method of analysis of field data

Survey data collected by multibeam echosounder Sea Beam 1180 of ELAC Nautik GmbH on r/v "Horyzont II" under IPY- Kinnvika expedition 2009. During work with the multibeam echosounder made continuous pinging and record of depths. The vessel followed route according to the voyage plan. In some cases the vessel deviated from planned route to avoid uncharted dangers or to collect more data of unknown area. The movement along the edge of the previously surveyed lane was also included into consideration.

The main aim of analysis was to find out correlations in between distance to the danger (the criterion of International Maritime Organization), depths and longitudinal and transverse changes of depths at the edge of surveyed lane. The analysed distances and bottom profiles (Fig.4) were not related to the coastline. Also they were not related to any isobaths. They were related to the non-linear movement and position of the vessel. This movement was the result of subjective assessment of the safety by the navigator. Position of the transducer of the multibeam echosounder was point of reference for depths and distances. The correlation represented by equations in between navigational safety parameters were received by PAST (Paleontological Statistics) software of Natural History Museum in Oslo.

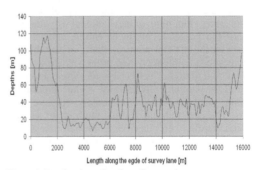

Figure 4. Depths along the edge of survey lane

3.2 Results

Analyzing series of data received during survey identified formula (Eqn 3) showing correlations in between distance to danger and longitudinal changes of depths.

$$D = 19.1484 - 0.2741 \ \Delta h_L \qquad (3)$$

where D – distance to the edge of the surveyed lane interpreted as distance to the danger (m); Δh_L – change of depths along the edge of the surveyed lane on the longitudinal section of 100 meters (m).

Same way identified formula (Eqn 4) showing correlations in between distance to the danger and transverse changes of depths.

$$d = 19.5455 - 0.8512 \; \Delta h_p \qquad (4)$$

where d – distance to the edge of the surveyed lane interpreted as distance to the danger (m); Δh_P – change of depths along the edge of the surveyed lane on transverse section of 100 meters (m).

4 CONCLUSIONS

The electronic charts made by various makers may be made in the different standard then accepted by the ship's ECDIS system being on board the vessel. In such case the ship-owner must solve dilemma of undertaking high buying costs of second ECDIS system that will serve for other electronic charts fulfilling necessities of the planned voyage. Purchase of next ECDIS system for a single or occasional voyage seems loose the financial competition with the paper or raster charts as far as such alternative exists.

The makers of the electronic charts being under pressure of strict requirements of ZOC are forced to downgrade quality of presented information on paper charts even for few groups. Issuing gratuitous unofficial charts (NOAA, http://charts.noaa.gov; http://www.nauticalcharts.noaa.gov, 16-Jan-10) or with considerably lower price than theirs official equivalents (Garmin Ltd, http://www8.garmin.com, 08-Jan-10; Jeppesen Norway A/S, http://www.c-map.no ,25-Jan-10) is favourable signal for sea charts users. Planning of navigational voyage support of the vessel seems to be simple and clear in case exist the official nautical charts of suitable parameters for the intended kind of the voyage.

The goal of author is to elaborate simple appraisal method for planning navigational voyage support including unsurveyed or inaccurately surveyed regions. The external and internal methods to support the navigation were described. Proper assessment scales and coefficients were proposed. Above method gave tool for appraisal of the navigational voyage plan in clear and simple manner so convenient for organizers and performers of a voyage. By this way is possible to detect the weaknesses of vessel's preparedness and improve it.

The unknown regions ("blank places") on screen of the multibeam echosounder are treated with distrust by the navigators. The distance to the edge of unsurveyed region that navigator try to hold is approximately 20 meters. This distance is inversely proportional to the tendency of changes. The angular dimensions of the image on screen (visual estimation of not processed image of the sea bottom relief), the range of beams of the echosounder and the distance to the edge of the surveyed lane are essential for estimation of safety. Identification of more detailed correlations requires however further research.

REFERENCES

Gale, H., 2009: From paper charts to ECDIS, A practical voyage plan: Part 3, Seaways November 2009, p. 5 - 9.
IHO, 1994: Hydrographic Dictionary, Part I, Volume I, English, Special Publication No. 32, Fifth Edition, Monaco 1994, 280 pp.
IHO, 2008: IHO Standards for Hydrographic Surveys, Special Publication No 44, Fifth Edition, Monaco February 2008, pp.28
IHO, 2009: Regulations of the IHO for international (INT) charts and charts specifications of the IHO, Edition 3.006, April 2009, 392 pp.
IMO, 1993: Resolution A.741(18), International Safety Management Code, 4 November 1993.
IMO, 2004: SOLAS consolidated edition 2004, London, pp.628.Jeppesen Norway A/S, 2010: C-MAP Chart Catalog v.2.4.0.8.
Pastusiak, T., 2010: Problems of coverage of unsurveyed regions by electronic charts [in Polish], Logistyka 2/1010, ISSN 1231–5478, p 2069-2086.
Primar Stavanger, 2009: Primar Chart Catalogue 4.3.
Statens Kartverk 2001: NO537, Paper chart, Svalbard, Hinlopenstretet N, scale 1:100,000.
The Norwegian Hydrographic Service and Norwegian Polar Research Institute, 1990: Den Norske Los, Arctic Pilot, Vol. 7, 2nd Edition, 433 pp.
Transas Marine Ltd, 2010: Transas Chart World Folio v.3.2.332, WF 28.
UKHO, 2004: NP100, The Mariners Handbook, 8th Edition, 260 pp.
UKHO, 2007: NP11 Arctic Pilot, Edition 2004, Correction 2007.
UKHO, 2010: Admiralty Digital Catalogue v.1.6.0.
UKHO, 2009: Catalogue of Admiralty Charts and Publications NP131, 2009 Edition.
Weintrit, A., 2009: The electronic chart display and information system (ECDIS) - An operational handbook, Boca-Raton - London - New York – Leiden, CRS Press, Taylor & Francis Group, ISBN-13: 978-0-41548246-2, 1101pp.

8. Methods of Iceberg Towing

A. Marchenko & K. Eik
The University Centre in Svalbard, Longyearbyen
STATOIL

ABSTRACT: Mathematical models of iceberg towing by a ship connected to the iceberg by mooring lines are considered. Governing equations describing the towing and the tension of mooring lines in two different schemes of the towing are formulated. Stability of steady solutions describing the towing with constant speed is studied. Numerical simulations are realized to compare results of the modeling with experimental results of model towing in HSVA ice tank.

1 INTRODUCTION

Icebergs may cause a threat to installations, vessels and operations in a number of Arctic and Antarctic regions. If icebergs are detected and considered to be a threat, it has been documented that they can be deflected into a safe direction in approximately 75% of the events (Rudkin et al., 2005). The preferred method for iceberg deflection is single vessel tow rope (Fig 1a). Experimental towing of the iceberg in the Barents Sea was realized with a rope taken around the iceberg by a loop in 2005, and the rope was broken during the towing. Photograph of the towing is shown in Fig. 1.

Figure 1. Towing of 200000 t iceberg in the Barents Sea.

The majority of the unsuccessful tows ended because the tow slipped over the iceberg while ruptures of tow line or iceberg rolling over are other common explanations (Rudkin et al., 2005). There are also examples of towing where towing in the planned direction was not possible. In order to in-

crease the understanding of what happens when an iceberg tow is started Marchenko and Gudoshnikov (2005) and Marchenko and Ulrich (2008) developed a numerical model for iceberg towing. Eik and Marchenko (2010) analysed results of HSVA experiments on the towing of model iceberg in open water and when broken ice was floating on the water surface. Stability of iceberg towing with floating rope was discussed in (Marchenko, 2010).

In the present paper we compare two different methods of iceberg towing, formulate governing equations, analyse the stability of steady towing and perform numerical simulations of the HSVA experiments on iceberg towing in open water taking into account the influence of natural oscillations of water in the tank.

2 METHODS OF ICEBERGS TOWING

In the practice of iceberg management two methods (I and II) of icebergs towing were performed. In both methods towed iceberg was trapped by a loop of floating synthetics rope ringed around the iceberg. The ends of the rope are fastened on the boat stern in the method I (Fig. 1a) and connected to heavy steel hawser fastened at the boat stern in the method II (Fig. 1b). The rope floats or hangs above the water surface in the method I. In the method II the hawser is submerged and the rope is floating near the iceberg and submerged near its connection to the hawser. Typically the hawser is shorter the rope. In Fig. 2 the line FTR shows floating tow rope and the line W shows the steel hawser. Thin line SS shows the water line around the iceberg and between the iceberg and the boat. In Fig. 2a) the rope segments D_1O_1 and

D_2O_2 are floating, and the segments O_1C and O_2C are hanging above water surface. In Fig. 2b) the rope segments O_1B and O_2B are submerged. The realization of the method I is simpler on the practice, but in case of the rope break up the boat stern can be damaged by the rope. In the method II the water takes buffer role in case of the rope break up.

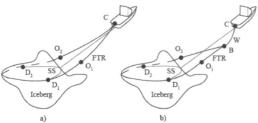

a)   b)

Figure 2. Schemes of iceberg towing with floating (a) (method I) and submerged (b) (method II) tow lines.

The forces applied to the iceberg ($\mathbf{F}_{Ir}$) and to the boat ($\mathbf{F}_{Br}$) by the rope in the method I are calculated with formulas

$$\mathbf{F}_{Ir} = \mathbf{T}_{D1} + \mathbf{T}_{D2}, \quad \mathbf{F}_{Br} = \mathbf{T}_{C1} + \mathbf{T}_{C2}. \qquad (1)$$

The rope tension forces $\mathbf{T}_{D1}$, $\mathbf{T}_{D2}$, $\mathbf{T}_{C1}$ and $\mathbf{T}_{C2}$ are shown in Fig. 3. The forces applied to the iceberg ($\mathbf{F}_{Ir}$) by the rope and to the boat ($\mathbf{F}_{Bh}$) by the hawser in the method II are calculated with formula

$$\mathbf{F}_{Ir} = \mathbf{T}_{D1} + \mathbf{T}_{D2}, \quad \mathbf{F}_{Bh} = \mathbf{T}_C. \qquad (2)$$

The rope tension forces $\mathbf{T}_{D1}$, $\mathbf{T}_{D2}$ and $\mathbf{T}_C$ are shown in Fig. 4.

Safety requires avoid the approaching of iceberg to the boat. Therefore in both methods of icebergs towing distances X_1 and X_2 shown in Fig. 3 and Fig. 4 should be much greater distances R_1 and R_2. It can be expressed by the inequality $\chi << 1$, where $\chi = R/X$ and R represents iceberg radius. Quantity X represents distance between the boat stern and the points D_1 and D_2, where the rope approaches to the iceberg. Assuming $X \approx 500$ m and representative radius of the iceberg 50 m we find that $\chi = 0.1$. Distances z_C and z_B are typically smaller 10 m. Therefore parameter $\varepsilon = z_{C,B}/l_r$ is smaller 0.02.

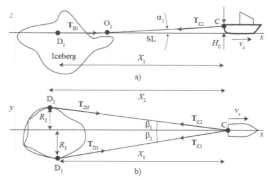

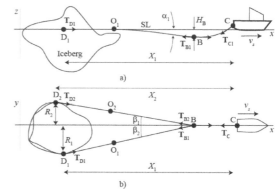

Figure 3. Schemes of towing line in method I in lateral (a) and upward (b) projections.

Figure 4. Schemes of towing line in method II in lateral (a) and upward (b) projections.

Angles $\alpha_{1,2}$ and $\beta_{1,2}$ shown in Fig. 3 and Fig. 4 are of the orders ε and χ respectively. Projections of the rope tensions on the x-direction are proportional to $\cos\alpha_{1,2}$ and $\cos\beta_{1,2}$. Therefore their difference has the order $O(\varepsilon^2, \chi^2)$, and formulas (1) and (2) can be written with accuracy $O(\varepsilon, \chi)$ as follows

$$F_{Ir} = T, \quad F_{Br} = -T, \qquad (1a)$$

$$F_{Ir} = T, \quad F_{Bh} = -T_C = -T, \qquad (2a)$$

where $T/2$ is the rope tension, F_{Ir}, F_{Br} and F_{Bh} are the absolute values of the forces applied to the iceberg and to the boat along the x-axis.

Further it is assumed that the difference between x-coordinates of the points D_1 and D_2 is much smaller than the iceberg radius R, and $X_1 = X_2 = X$ with accuracy $O(\varepsilon, \chi)$. In this case the rope tension is expressed as a function

$$T = T(X), \quad dT/dX > 0 \qquad (3)$$

where X is the distance between points of the rope fastening on the boat stern and on the iceberg surface. Inequality in (3) means that the increase of X is accompanied by the increase of the rope tension.

3 GOVERNING EQUATIONS

Equations of momentum balance for boat and iceberg connected by tow line are written as follows

$$M_s \frac{dv_s}{dt} = -R_s - T + P - M_s g \partial h / \partial x, \qquad (4)$$

$$M_i \frac{dv_i}{dt} = -R_i + T - M_i g \partial h / \partial x, \qquad (5)$$

where M_s and M_i are the masses of the boat and the iceberg including the added mass, P is the boat propulsion, and $\partial h / \partial x$ is the water surface gradient. It is assumed that horizontal scale of water surface elevation is much greater than the iceberg diameter and ship length. Water resistances to boat motion (R_s) and to iceberg motion (R_i) are described by square low as follows

$$R_s = \rho_w C_{ws} S_s |v_s - u|(v_s - u), \ R_i = \rho_w C_{wi} S_i |v_i - u|(v_i - u), \quad (6)$$

where S_s is wet surface of the boat hull, and S_i is representative area of vertical cross-section of submerged surface of the iceberg in perpendicular direction to the tow direction, and u is water velocity. Drag coefficients are equal to $C_{ws} = 0.003$ (Voitkunsky, 1988) and $C_{wi} \in (0.5,1)$ (Robe, 1980).

Equations (4) and (5) are completed by the definition of relative velocity as follows

$$\frac{dX}{dt} = v_s - v_i. \qquad (7)$$

Equations (4), (5) and (7) perform closed system of ordinary differential equations relatively unknown functions of the time $v_s(t)$, $v_i(t)$ and $X(t)$. They have steady solution describing steady towing with constant speed v_0, constant propulsion P_0, constant water velocity u and $\partial h / \partial x = 0$. For the steady towing it follows

$$v_s = v_i = v_0, \ T_0 = R_i, \ R_s + T_0 = P_0, \qquad (8)$$

$$v_0 - u = \sqrt{\frac{P_0}{\rho_w (C_{ws} S_s + C_{wi} S_i)}}, \ T_0 = (v_0 - u)^2 \rho_w C_{wi} S_i \quad (9)$$

Distance $X = X_0$ is determined from the first formula (3) when $T = T_0$.

Dimensionless variables: time t', velocities v'_s and v'_i, rope tension τ, boat propulsion π and dis-

tance Z between points of the rope fastening are introduced as follows

$$t' = \frac{t}{t_{r,1}}, \ v'_s = \frac{v_s}{v_0}, \ v'_i = \frac{v_i}{v_0}, \ u' = \frac{u}{v_0}, \ \tau = \frac{T}{T_0},$$

$$\pi = \frac{P}{P_0}, \ Z = \frac{X}{l_r}, \qquad (10)$$

where $t_{r,1} = M_s v_0 / T_0$ is the representative time, and l_r is the rope length. Equations (4), (5) and (7) are written in dimensionless variables as follows (primes near dimensionless variables are omitted)

$$\frac{dv_s}{dt} = -\varepsilon_1 |v_s - u|(v_s - u) - \tau + (1 + \varepsilon_1)\pi - \varepsilon_w \partial \eta / \partial x, \quad (11)$$

$$\frac{dv_i}{dt} = -\varepsilon_2 |v_i - u|(v_i - u) + \varepsilon_2 \tau - \varepsilon_w \partial \eta / \partial x, \ \gamma \frac{dZ}{dt} = v_s - v_i,$$

where $\varepsilon_1 = C_{ws} S_s / (C_{wi} S_i)$, $\varepsilon_2 = M_s / M_i$, $\gamma = l_r T_0 / (M_s v_0^2)$, $\varepsilon_w = a / l_r$, and a is the amplitude of water surface elevation. Simple estimates show that $\varepsilon_{1,2} \ll 1$ when the iceberg is not very small.

Steady solution in dimensionless variables is written as

$$v_s = v_i = \tau = \pi = 1. \qquad (12)$$

Value $Z = Z_0$ for steady towing is constructed as solution of the equation $\tau(Z_0) = 1$.

4 STABILITY OF STEADY TOWING

Solution of equations (11) in the vicinity of the steady point is written in the form

$$v_s = 1 + \delta v_s, \ v_i = 1 + \delta v_i, \ Z = Z_0 + \delta Z, \qquad (13)$$

where new independent variables $\delta v_s(t)$, $\delta v_i(t)$ and $\delta \Sigma(t)$ describe small fluctuations in the vicinity of the steady solution. Substitution of formulas (13) in equations (11) leads to the following equations

$$\frac{d\delta v_s}{dt} = -2\varepsilon_1 \delta v_s - \tau_0' \delta Z, \ \frac{d\delta v_i}{dt} = -2\varepsilon_2 \delta v_i + \varepsilon_2 \tau_0' \delta Z,$$

$$\gamma \frac{d\delta Z}{dt} = \delta v_s - \delta v_i, \qquad (14)$$

where $\tau_0' = d\tau / dZ$ by $Z = Z_0$.

Substituting exponential solution $\delta v_s = \delta v_{s,0} e^{\psi t}$, $\delta v_i = \delta v_{i,0} e^{\psi t}$, $\delta Z = \delta Z_0 e^{\psi t}$, in equations (14) we find that eigenvalues ψ are the roots of the cubic equation

$$F(\psi) \equiv \psi(\psi + 2\varepsilon_1)(\psi + 2\varepsilon_2) + \mu(\psi(1 + \varepsilon_2) + 2\varepsilon_2(\varepsilon_1 + \varepsilon_2)) = 0, \ (15)$$

where $\mu = \tau_0'/\gamma > 0$. Roots $\psi_{i,0}$ of equation (15) by $\varepsilon_1 = \varepsilon_2 = 0$ are chosen as the first approximation of the roots when $\varepsilon_{1,2} \ll 1$. One finds

$$\psi_{1,0} = 0, \quad \psi_{2,0} = i\sqrt{\mu}, \quad \psi_{3,0} = -i\sqrt{\mu}. \qquad (16)$$

Next approximation is expressed by the formulas

$$\psi_1 = -2\varepsilon_2, \quad \psi_2 = i\sqrt{\mu} - \varepsilon_1 + \frac{i\sqrt{\mu}}{2}\varepsilon_2,$$

$$\psi_3 = -i\sqrt{\mu} - \varepsilon_1 - \frac{i\sqrt{\mu}}{2}\varepsilon_2. \qquad (17)$$

From formulas (17) follows that real parts of eigen-values ψ are negative. Therefore the solution describing steady towing is stable. At the same time absolute values the of real parts of eigen-values ψ_2 and ψ_3 are much smaller than absolute values of their imaginary parts: $|\mathrm{Re}(\psi_{2,3})| \ll |\mathrm{Im}(\psi_{2,3})|$. Therefore damping of perturbations of the steady solution will be accompanied by oscillations with dimensionless frequency $\sqrt{\mu}$. Dimensional period of the oscillations is calculated as follows $t_{p,1} = 2\pi t_{r,1}/\sqrt{\mu}$. Oscillations of mooring system with floating submerged sensors were observed and studied by Hamilton (2000).

5 MODEL TESTS OF ICEBERGS TOWING

Model tests on the iceberg towing were performed at Hamburg Ship Model Basin (HSVA), Germany. Two model icebergs with cylindrical and rectangular shapes were made from water ice and towed using scheme shown in Fig. 1b) in the tank with different concentration of ice floes on the water surface. The description and results of the experimental studies are performed in the paper (Eik and Marchenko, 2010). In the present paper we consider only results of the experiment when cylindrical iceberg was towed in the water with free surface. Dimensions of model iceberg and fragment of the towing are shown in Fig. 5.

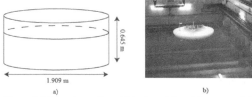

a)   b)

0.645 m

1.909 m

Figure 5. Dimensions of model iceberg (a) and fragment of the HSVA experiment (b).

The movement and rotation of towed icebergs were recorded in all six degrees of freedom with a Qualisys-Motion Capture System. The platform with sensors was installed on the iceberg surface to monitor the iceberg motion as it is visible in Fig. 5b). Three degrees of the movement are characterized by the horizontal distance between the carriage and the sensors (surge), sideways movement between fixed point at the carriage and the sensors (sway), and by the vertical displacement of the sensors (heave). Three degrees of the rotation are performed by the pitch, the roll and the yaw of the platform with sensors with respect to the direction of the tank extension.

The tow line consisted of floating rope Dyneema and steel wire. Characteristics of the towing rope Dyneema and the wire used in the tests are performed in Tables 1 and 2 for model scale. The tension in the tow line was recorded in three locations: on the end of the wire on the carriage and on the rope ends near the point of their connection with the wire (point B in Fig. 5). Some of the loads cells in some of the tests were "drifting" and manually corrected. This causes some unfortunate uncertainties in the load results. Average tow loads were varied in the range from 0.78 N to 4 N in the tests with open water.

The length and the water depth in the tank are equal $L_t = 72\,\mathrm{m}$ and $H_t = 2.5\,\mathrm{m}$ respectively (Fig. 5). Natural oscillations of the water in the tank can influence the towing. Horizontal water velocity u and water surface elevation h of the first natural mode are described by formulas

$$u = u_0 \cos\omega_1 t \sin k_{1x}x, \quad h = h_0 \sin\omega_1 t \cos k_{1x}x, \qquad (18)$$

where t is the time, x is the horizontal coordinate directed along the tank, h_0 is the amplitude of water surface elevation and $u_0 = -\omega_1 h_0/(k_{1x}H_t)$ is the amplitude of water velocity oscillations in the first mode, $k_{1x} = \pi/L_t$ is the wave number, and $\omega_1 = \sqrt{gH_t}k_1$ is the wave frequency. The period T_1 of the first natural mode is calculated with formula

$$T_1 = 2\pi/\omega_1 \approx 29\,\mathrm{s}. \qquad (19)$$

Water surface deformed by the first natural mode of the tank is shown in Fig. 5 by blue dashed and continuous lines at the different phases of the oscillation. Fig. 5 also explains that the distance X between the center of the iceberg and the carriage can be different from the distance X_{cs} between the sensor platform and the carriage because of the natural oscillation of the water in the tank and the iceberg pitch and roll. Difference $X - X_{cs}$ can depend on the sway and the yaw of the iceberg if the position of the sensor platform relatively the iceberg center is determined with insufficient accuracy.

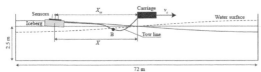

Figure 6. Scheme of the towing in the HSVA tank.

Period of heave oscillations of floating ice cylinder is estimated with the formula

$$T_c = 2\pi \left(\rho_w h_c / (\rho_w - \rho_i) g \right)^{1/2}, \qquad (20)$$

where $\rho_w = 1000\,\text{kg/m}^3$ and $\rho_i = 887\,\text{kg/m}^3$ are water and ice densities, and h_c is the cylinder height. Assuming $h_c = 0.645\,\text{m}$ we find the period of heave oscillation $T_c = 4.8\,\text{s}$ for the iceberg model.

The speed of the carriage in the experiment is shown in Fig. 7a) versus the time Mean tow speed is equal to 0.11 m/s when $125\,\text{s} < t < 250\,\text{s}$, and it is 0.13 m/s when $270\,\text{s} < t < 400\,\text{s}$. t. Fig. 8a) shows the difference between the carriage speed and its moving average calculated over 6 s. The mean deviation of the carriage speed is about 0.005 m/s. The carriage speed doesn't include oscillations with well recognized periodicity. Fig. 7b,c,d) show the surge, sway and heave of the iceberg versus the time. The characteristics of iceberg rotation performed by the pitch, roll and yaw are shown in Fig. 8b,c,d). The surge, sway, roll and yaw have visible correlation due to long term trend changing its direction when $t \approx 200\,\text{s}$. The surge and heave have oscillations with period about 30 s closed to the period T_1 of the first natural mode of the tank.

The amplitude of the heave oscillations varied within 0.5 - 1 cm when the iceberg was near the edge of the tank and decreases to smaller values when the iceberg was towed to the center of the tank. From the first formula (18) it follows that the amplitude of water velocity oscillations in the first mode u_0 is varied from 2 cm/s to 4 cm/s in the middle part of the tank when $h_0 = 0.5 - 1\,\text{cm}$. The decrease of the heave amplitude in Fig. 6d) with the time can relate to the decrease of the amplitude of water surface elevation in the middle part of tank according to the second formula (18).

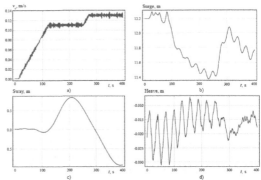

Figure 7. The speed of the carriage (a), horizontal distance between the carriage and the sensors installed on the iceberg (b), sideways movement relative between fixed points at carriage and the sensors (c) and iceberg heave (d) versus the time.

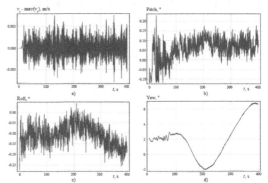

Figure 8. Difference between the carriage speed and its moving average (a), pitch (b), roll (c) and yaw (d) of the iceberg versus the time.

6 RESULTS OF NUMERICAL SIMULATIONS

Characteristics of towing lines in steady towing. Numerical simulations are performed in model and full scales. Properties of the towing rope and the hawser are performed in Tables 1 and 2. Towing rope Dyneema was used in the HSVA tests. Characteristics of the towing rope Dyneema performed in Table 1 for the full scale are taken from the web-address www.dynamica-ropes.dk. It is assumed that the diameter of the hawser should be the same as the rope diameter since their strengths are almost same.

Table 1. Tow rope properties (model and full scales)

Property	Unit	Model scale	Full scale
Total length	[m]	23	920
Length CD (Fig. 2)	[m]	10	400
Length BD (Fig. 3)	[m]	10	400
Weight	[Kg/m]	0.0069	6.22
Diameter	[m]	0.004	0.12
E-Module	[GPa]	95	95
Ultimate Load	[T]	1.3	1000

Table 2. Tow hawser properties (model and full scales)

Property	Unit	Model scale	Full scale
Total length	[m]	2.05	82
Weight	[Kg/m]	0.049	79.6
Diameter	[m]	0.003	0.12
E-Module	[GPa]	200	200
Ultimate Load	[T]	-	1000

From formulas (1a) and (2a) it follows that the towing with rope loop around the iceberg can be performed as the towing with one rope having double weight $2W_r$ and double buoyancy force $2W_b$. Models of towing lines related to towing schemes shown in Fig. 1 are performed in the Appendix. Fig. 9 perform the dependence of the rope tension T from the distance $\Delta X = l_{TL} - X$, where l_{TL} is total length of the towing line between the points C and D, in model and full scales. Curves 1 and 2 are related to towing method I and II shown in Fig. 1a) and Fig. 1b) respectively, $l_{TL} = 12.05$ m in model scale and $l_{TL} = 482$ m in full scale. One can see that the slope of the tension curve for the first towing scheme is higher than for the second towing scheme when the tension is relatively high.

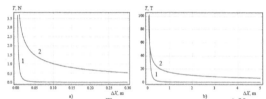

Figure 9. Rope tension T versus distance decreasing ΔX between the boat and the iceberg in model (a) and full (b) scales.

The shape of towing lines in steady towing performed by schemes shown in Fig.1a) and Fig. 1b) are performed in Fig. 10a,b) and Fig. 10c,d) respectively in model and full scales. One can see that all curves in Fig. 10 have small slopes. Thus assumptions made in the Appendix for the construction of the towing lines models are satisfied. It is also visible that towing rope always has floating part. In this case variations of the rope tension influence significantly vertical displacement of the towing lines and have insignificant influence on horizontal displacements of points C, D and B in comparison with the rope length. These displacements are invisible in Fig. 10 and therefore are shown by squares.

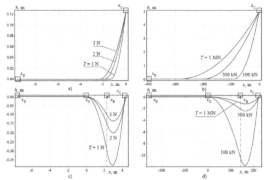

Figure10. The shape of towing lines in steady towing in model (a,c) and full scales (b,d) calculated with different values of the rope tension T.

Oscillations in unsteady towing. Dimensional period of these oscillations is estimated by the formula $t_p = 2\pi t_* / \sqrt{\mu}$, where $t_{r,2} = M_I v_0 / T_0$. Periods t_p are shown in Fig. 11 versus the tow load T in model and full scales. Curves 1 and 2 are related to the towing schemes shown respectively in Fig. 1a) and Fig. 1b). The periods are decreased with the increasing of the tow load. Representative value of the period is few tens of seconds when the towing is occurred according to Fig. 1b). Periods of oscillations in the towing scheme in Fig. 1a) are smaller 10 sec in model scale and 20 sec in full scale when tow load is relatively high. Periods of oscillations in the towing scheme in Fig. 1b) are higher then in the towing scheme in Fig. 1a) . From Fig. 11a) follows that period t_p is closed to period T_1 of the first natural mode of the tank, when the rope tension is about 2 N. It can create resonance effect. In full scale period t_p can be closed to swell period. For the conditions of the Barents Sea swell period is about 12 s. From Fig. 10b) it follows that the swell period can be closed to the period T_1 when the towing is performed according to the scheme in Fig. 1a).

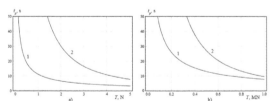

Figure 11. Periods of oscillations in model (a) and full (b) scales.

Fig. 12a,c) show the data measured in the experiment on the towing of model iceberg with cylindrical shape in the HSVA tank (Eik and Marchenko, 2010). Iceberg diameter and height are 1.909 m and

0.645 m. The towing was realized as it is shown in Fig. 1b). Characteristics of tow line are shown in Tables 1, 2. The rate of the distance between the carriage and iceberg (dX/dt) is calculated using the experimental data. The iceberg motion was calculated using equations (11), where v_s was substituted equaling to the carriage velocity $v_c(t)$. Initial conditions were $v_i = 0$ and $Z = 1.19$ by $t = 0$. It is assumed that $v_0 = 0.11\,\mathrm{m/s}$, $M_i = 1643\,\mathrm{kg}$ and $T_0 = 2.34\,\mathrm{N}$. Figure 13a) shows carriage velocity used in numerical simulations. Computed displacement of the iceberg is shown in Fig. 13b). Fig. 13a) shows surge rate dX/dt calculated using experimental data. Fig. 13b) shows the surge rate without accounting of the natural oscillations of the water in the tank. Figures 13c) and Fig. 13d) shows the surge rate with accounting of the natural oscillations with period 29 s and 16 s respectively. One can see that periods and amplitudes of oscillations of dX/dt in Fig. 13a) are most close to those performed in Fig. 13c).

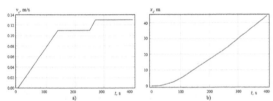

Figure 12. Carriage velocity versus the time used in simulations (a). Calculated iceberg displacement versus the time (b).

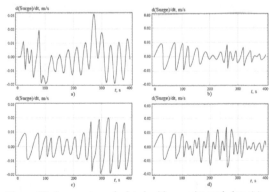

Figure 13. Surge rates calculated with experimental data (a), with numerical simulations without accounting of natural oscillations of the water in the tank (b) and with accounting of natural oscillations of the water in the tank with periods 29 s (c) and 16 s (d).

7 CONCLUSIONS

Stability of steady towing of iceberg is analyzed for two methods of the towing (Fig. 2). It is shown that steady towing is stable, but damping of steady towing perturbations is accompanied by oscillations in the system ship-tow line-iceberg. The resonance of these oscillations with swell of 12 s period observed in the Barents Sea is available when the towing is realized with floating tow line (method I). Model experiments on iceberg towing were performed in HSVA ice tank. Surge and heave oscillations of model icebergs were observed in the experiments. Periods of these oscillations were close to the period of the first natural mode of water oscillations in the tank 29 s. Numerical simulations confirm that the amplitudes of surge and heave oscillations of iceberg models were influenced by the natural oscillations of the water in the tank.

REFERENCES

Eik, K., Marchenko, A., 2010. Model tests of iceberg towing. Cold Regions Science and Technology, 61, pp. 13-28.

Hamilton, J.M., 2000. Vibration-based technique for measuring the elastic properties of ropes and the added masses of submerged objects. J. of Atmospheric and Oceanic Technology. Vol. 17, pp. 688-697.

Marchenko, A., Gudoshnikov, Yu., 2005. The Influence of Surface Waves on Rope Tension by Iceberg Towing. Proc. of 18th Int. Conference on Port and Ocean Engineering under Arctic Conditions (POAC'05), Vol. 2, Clarkson University, Potsdam, NY, pp.543-553.

Marchenko, A., and C., Ulrich, 2008. Iceberg towing: analysis af field experiments and numerical simulations. Proceedings of 19th IAHR International Symposium on Ice "Using New Technology to Understand Water-Ice Interaction", Vancouver, BC, Canada, July 6-11, 2008, Vol.2, 909-923.

Marchenko, A.V., 2010. Stability of icebergs towing. Transactions of the Krylov Shipbuilding Research Institute. Marine Ice Technology Issue. 51 (335), 69-82. ISBN 0869-8422. (in Russian)

McClintock, J., McKenna, R. and Woodworth-Lynas, C. (2007). Grand Banks Iceberg Management. PERD/CHC Report 20-84, Report prepared by AMEC Earth & Environmental, St. John's, NL, R.F. McKenna & Associates, Wakefield, QC, and PETRA International Ltd., Cupids, NL., 84 p.

Robe, R.Q., 1980. Iceberg drift and deterioration. In: Colbeck, S. (Ed.), Dynamics of Snow and Ice Masses. Academic Press, New York. pp. 211-259.

Rudkin, P., Boldrick, C. and Barron Jr., P., 2005. PERD Iceberg Management Database. PERD/CHC Report 20-72, Report prepared by Provincial Aerospace Environmental Services (PAL), St. John's, NL., 71 p.

Voitkunsky J.I., 1988. Resistance to ship motion. Leningrad: Sudostroenie. 287 p. (in Russian)

APPENDIX

Model of the towing with floating rope

The scheme of icebergs towing with floating rope is shown in Fig. 14a). The rope segment OC hangs in the air, and the rope segment OD floats on the water surface. The coordinate of the point O is equal to $-x_0$, and the coordinate of point D is equal to $-X$. The rope is fastened to the boat stern in the point C with coordinates $x=0$ and $z=z_C$. The motion of the rope is occurred under the influence of the gravity force, the rope tension and the rope inertia. Momentum balance of the rope segment hanging in the air with length ds is written as follows

$$\frac{W}{g}\mathbf{a} = \frac{d}{ds}\mathbf{T} + \mathbf{W}, \qquad (\Pi 1)$$

where $\mathbf{W} = (0, -W_r)$ is the weight of the rope of unit length, $\mathbf{T}$ is the rope tension, $\mathbf{a}$ is the acceleration (Fig. 14c). Since the vector of the rope tension is tangential to the rope it can be performed as $\mathbf{T} = T\boldsymbol{\tau}$, where $\boldsymbol{\tau}$ is the tangential vector of unit length to the rope segment ds. Using Frenet formulas equation ($\Pi 1$) is written in the form

$$\frac{W}{g}\mathbf{a} = \frac{dT}{ds}\boldsymbol{\tau} - Tk\mathbf{n} + \mathbf{W}, \qquad (\Pi 2)$$

where $\mathbf{n}$ is unit normal vector to the rope segment ds and k is the rope curvature.

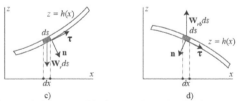

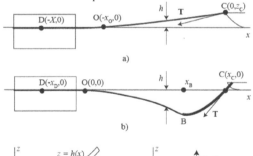

Figure 14. Schemes of iceberg towing with floating (a) and submerged (b) ropes. Schemes of forces applied to the rope hanging in the air (c) and to submerged rope (b).

The shape of the hanging rope is described by equation $z = h(x,t)$, where x and z are the horizontal and vertical coordinates and t is the time. The rope curvature is calculated by the formula

$k = \left(\partial^2 h/\partial x^2\right)\left(1 + \left(\partial h/\partial x\right)^2\right)^{-3/2}$. In static case the integration of the projection of equation ($\Pi 2$) on tangential vector $\boldsymbol{\tau}$ leads to the formula $T = T_{st} + W_r h$, where T_{st} is a constant. The projection of equation ($\Pi 2$) on normal vector $\mathbf{n}$ is written as follows

$$T\frac{d^2 h}{dx^2} = W_r\sqrt{1 + \left(\frac{dh}{dx}\right)^2}, \qquad (\Pi 3)$$

Dimensionless variables are introduced by formulas

$$\varsigma = \frac{x}{l_{r,0}}, \quad \tau = \frac{T_{st}}{T_0}, \quad \eta = \frac{h}{z_C}, \qquad (\Pi 4)$$

where $l_{r,0}$ is the rope length and T_0 is the rope tension in steady towing. In dimensionless variables equation ($\Pi 3$) has the following form

$$(\tau + \upsilon\eta)\frac{\varepsilon}{\upsilon}\frac{d^2\eta}{d\varsigma^2} = \sqrt{1 + \left(\varepsilon\frac{d\eta}{d\varsigma}\right)^2}, \qquad (\Pi 5)$$

where dimensionless coefficients ε and υ are introduced as follows

$$\varepsilon = \frac{z_C}{l_r}, \quad \upsilon = \frac{W_r z_C}{T_0}. \qquad (\Pi 6)$$

The rope tension T_0 is estimated with formula (8). Assuming $z_C = 10\,\text{m}$, $W_r = 50\,\text{Nm}^{-1}$, $S_0 = 10^3\,\text{m}^2$, $C_{wi} = 0.6$ and $v_0 = 0.5\,\text{ms}^{-1}$ we find that $T_0 = 150\,\text{kN}$ and $\upsilon \approx 0.033$.

With accuracy up to high order terms equation ($\Pi 3$) is written in dimensionless variables ($\Pi 4$) as follows

$$\frac{d^2\eta}{d\varsigma^2} = \frac{k_r}{\varepsilon\tau}, \quad \varsigma \in (-\varsigma_0, 0). \qquad (\Pi 7)$$

where $k_r = l_r W_r / T_0$ and τ is dimensionless rope tension.

Boundary conditions for equation ($\Pi 7$) are formulated as follows

$$\eta = 1, \quad \varsigma = 0, \qquad (\Pi 8)$$

$$\eta = 0, \quad \frac{d\eta}{d\varsigma} = 0, \quad \varsigma = -\varsigma_0. \qquad (\Pi 9)$$

From ($\Pi 8$) and ($\Pi 9$) it follows

$$\eta = \frac{k_r\varsigma}{\varepsilon\tau}\left(\frac{\varsigma}{2} + \varsigma_0\right) + 1, \quad \varsigma \in (-\varsigma_0, 0), \qquad (\Pi 10)$$

$$\tau = \frac{k_r\varsigma_0^2}{2\varepsilon}. \qquad (\Pi 11)$$

Dimensionless length of the rope segment hinging in the air is calculated from the formula

$$l_{OC} = \int\limits_{-\varsigma_0}^{0} \sqrt{1+\varepsilon^2\left(\frac{d\eta}{d\varsigma}\right)^2}\, d\varsigma = \int\limits_{-\varsigma_0}^{0} \left(1+\frac{\varepsilon^2}{2}\left(\frac{d\eta}{d\varsigma}\right)^2\right) d\varsigma + O(\varepsilon^4), \quad (\Pi12)$$

and total length of the rope is equal to

$$1 = l_{OC} + Z - \varsigma_0. \qquad (\Pi13)$$

From formula ($\Pi12$) it follows

$$l_{OC} = \varsigma_0\left(1+\frac{2\varepsilon^2}{3\varsigma_0^2}\right). \qquad (\Pi14)$$

From formulas ($\Pi13$) and ($\Pi14$) follows

$$\varsigma_0 = \frac{2\varepsilon^2}{3(1-Z)}. \qquad (\Pi15)$$

Substituting ($\Pi15$) in formula ($\Pi11$) we find the expression of the rope tension

$$\tau = \frac{2k_r\varepsilon^3}{9(1-Z)^2}. \qquad (\Pi16)$$

In the steady towing $\tau = 1$ and $Z_0 = 1-\sqrt{2\varepsilon k_r}/3$. Dimensionless parameter $\mu = \tau_0'/\gamma$ is calculated with the formula

$$\mu = \frac{3\sqrt{2}}{\gamma\varepsilon\sqrt{\varepsilon k_r}}. \qquad (\Pi17)$$

Formula ($\Pi16$) is applicable when $\varsigma_0 \le Z$. This condition imposes limitation for the tension $\tau \le \tau_{cr}$, where $\tau_{cr} = k_r/(2\varepsilon^2)$. The rope hangs above the water when $\tau > \tau_{cr}$. Dimensional value of critical rope tension $T_{cr} = l_r W_r/(2\varepsilon^2) \approx 1632\,\text{T}$, when $l_r = 400\,\text{m}$, $z_C = 10\,\text{m}$, $W_r = 50\,\text{Nm}^{-1}$. This value is much greater the strength of synthetic ropes.

Model of the towing with submerged rope

Scheme of iceberg towing with submerged rope is shown in Fig. 14b). The rope is fastened on iceberg surface behind the point D, it is floating between points D and O and submerged between points O and B. The rope BD is connected to the wire BC fastened at the boat stern. The length of the rope between the points C and D is equal to l_r, and the length of the wire between the points B and C is equal to l_w. It is convenient to set up the origin in the point O. The coordinates of the points D, O, B and C are equal to $(-x_D,0)$, $(0,0)$, $(x_B,-z_B)$ and $(x_C,0)$ respectively. It is assumed that the weight of the rope with unit length is equal to W_r, and it is smaller than

the buoyancy force W_b applied to the rope. Therefore the resulting force $W_{rb} = W_b - W_r$ is upward directed (Fig. 14d). The weight of the wire W_w with unit length is much greater the buoyancy force applied to the wire, therefore the influence of the buoyancy force is ignored.

Analogically ($\Pi7$) the shape of submerged rope OB is described in dimensionless variables by the equation

$$\frac{d^2\eta}{d\varsigma^2} = -\frac{k_{rb}}{\tau}, \quad \varsigma \in (0,\varsigma_B), \qquad (\Pi18)$$

where $k_{rb} = l_r W_{rb}/T_0$ and τ is dimensionless tension of the rope and the wire. Since z_B is unknown quantity depending on the solution dimensionless variable η is determined by formula $\eta = h/l_r$ in contrast with last formula ($\Pi4$). Nevertheless further it is assumed that $|d\eta/d\varsigma| \ll 1$.

Boundary conditions in the point O for the construction of the solution of equation ($\Pi18$) describing the shape of the rope BD are formulated as follows

$$\frac{d\eta}{d\varsigma} = \eta = 0, \quad \varsigma = 0 \qquad (\Pi19)$$

From ($\Pi18$) and ($\Pi19$) it follows

$$\eta = -\frac{k_{rb}}{2\tau}\varsigma^2, \quad \varsigma \in (0,\varsigma_B). \qquad (\Pi20)$$

The shape of the wire BC is described by the equation

$$\frac{d^2\eta}{d\varsigma^2} = \frac{k_w}{\tau}, \quad \varsigma \in (\varsigma_B,\varsigma_C), \qquad (\Pi21)$$

where $k_w = l_r W_w/T_0$. The solution of equation ($\Pi21$) satisfying to the condition $\eta = 0$ by $\varsigma = \varsigma_C$ is written as follows

$$\eta = \frac{k_w}{2\tau}(\varsigma - \varsigma_C)(\varsigma + B). \qquad (\Pi22)$$

Matching conditions for solutions ($\Pi20$) and ($\Pi22$) have the form

$$\lim_{\varsigma\to\varsigma_B-0}\eta = \lim_{\varsigma\to\varsigma_B+0}\eta, \quad \lim_{\varsigma\to\varsigma_B-0}\frac{d\eta}{d\varsigma} = \lim_{\varsigma\to\varsigma_B+0}\frac{d\eta}{d\varsigma}. \qquad (\Pi23)$$

Analogically ($\Pi12$) dimensionless unit length of the rope between the points D and B and the wire length between the points B and C are expressed by formulas

$$l_{DB} = \int_{-\varsigma_D}^{\varsigma_B} \left(1 + \frac{1}{2}\left(\frac{d\eta}{d\varsigma}\right)^2\right) d\varsigma = 1,$$

$$l_{BC} = \int_{\varsigma_B}^{\varsigma_C} \left(1 + \frac{1}{2}\left(\frac{d\eta}{d\varsigma}\right)^2\right) d\varsigma = \lambda_{wr}, \tag{П24}$$

Where dimensionless length of the wire in equal to the ratio or dimensional lengths of the wire and the rope (l_w and l_r): $\lambda_{wr} = l_w / l_r$.

Dimensionless distance between the boat and the iceberg is introduced by the formula

$$Z = \varsigma_D + \varsigma_C. \tag{П25}$$

Formulas (П20), (П22) and conditions (П23) - (П25) are used to perform implicit dependence between dimensionless rope and wire tension τ and distance Z. After simple algebra the implicit dependence is reduced to algebraic equations

$$\beta\varsigma_C + \frac{\alpha\varsigma_C^3}{\tau^2} = \lambda_{wr}, \quad Z - \beta\varsigma_C + \frac{\delta\varsigma_C^3}{\tau^2} = 1, \tag{П26}$$

where coefficients α, β and δ are expressed by formulas

$$\alpha = \frac{\beta k_w^2}{6}\left(\beta^2 + 3(1-\beta)^2\left(\frac{k_{rb}}{k_w}\right)^2 - 3\beta(1-\beta)\frac{k_{rb}}{k_w}\right),$$

$$\beta = \sqrt{\frac{k_{rb}}{k_{rb} + k_w}}, \quad \delta = \frac{k_{rb}^2(1-\beta)^3}{6}. \tag{П27}$$

Excluding ς_C from equations (П26) we find explicit formula for the rope tension

$$\tau = \sqrt{\frac{\alpha\varsigma_C^3}{\lambda_{wr} - \beta\varsigma_C}}, \quad \varsigma_C = \frac{\lambda_{wr}\delta - \alpha(1-Z)}{\beta(\alpha+\delta)}. \tag{П28}$$

Tension $\tau \to \infty$ when $Z = 1 + \lambda_{wr}$ or when $X = l_r + l_w$ in dimensional variables.

In steady solutions $\tau = 1$ and rope characteristics in steady towing ς_C^0 and Z_0 are determined as follows

$$\beta\varsigma_C^0 + \alpha\left(\varsigma_C^0\right)^3 = \lambda_{wr}, \quad Z_0 = 1 + \beta\varsigma_C^0 - \delta\left(\varsigma_C^0\right)^3. \tag{П29}$$

Taking differential from equations (П26) in the vicinity of $\tau = 1$ we find

$$\tau_0' = \frac{\beta + 3\alpha\left(\varsigma_C^0\right)^2}{2\beta(\alpha+\delta)\left(\varsigma_C^0\right)^3}. \tag{П30}$$

Dimensionless parameter μ characterizing the stability of steady towing is calculated with formula $\mu = \tau_0' / \gamma$.

9. Ice Management – From the Concept to Realization

I.Ye. Frolov, Ye.U. Mironov, G.K. Zubakin, Yu.P. Gudoshnikov, A.V. Yulin,
V.G. Smirnov & I.V. Buzin
Arctic and Antarctic Research Institute, Saint-Petersburg, Russia

ABSTRACT: In the present time in the Russian Arctic and freezing seas there's the growth of industrial activity. In addition to the traditional navigation in the ice-infested waters the development of the new offshore hydrocarbon deposits is planned. New production and uploading platforms and high-tonnage tankers appear in the Arctic. Widening of the sea activities in the Arctic, the implementation of comprehensive technical projects and the need to ensure their safety made it necessary to develop and introduce principally new information and logistics system - a system aimed at "managing ice conditions" or the so-called "Ice Management" (IM). The vast experience of the informational support of the ice navigation is accumulated in Russia, many components of the IM are developed and implemented in the active practice. The paper presents the summary of such experience. The concept of development the IM on the Shtokman Gas Condensate Field is discussed.

1 PURPOSE OF ICE MANAGEMENT

In order to provide active navigation, countries having access to the Arctic seas elaborated systems of hydrometeorological support (HMS). This term means set of actions aimed at systematic support of subjects carrying out activities in the Arctic with information on current and foreseeable ice and hydrometeorological conditions along the navigational routes or in the regions of production activity.

Widening of the sea activities in the Arctic, the implementation of comprehensive technical projects and the need to ensure their safety made it necessary to develop and introduce principally new information and logistics system - a system aimed at "at managing ice conditions" or the so-called "Ice Management" (IM). "Ice Management" is the system of specific actions for advance assessment and early prevention of dangerous ice phenomena, including as well active use of the technical means to effect the ice formations in order to reduce their negative impact on the offshore producing facilities

In contrast to HMS system, which is simply informational system, IM could be referred to as a group of comprehensive information and logistics systems. The main purpose of these systems is to assess risk of arising situation in advance and develop recommendations on methods of active management of the situation.

2 INFORMATIONAL SYSTEM OF RUSSIA ON SEA ICE

Arrangement of the ice management is based on ice and informational systems, which by now have been developed in the majority of national ice services. Creation and development of these services is predicated on their tasks on informational support of various activities in the freezing water areas. Any ice informational system consists of three main blocks: one of the collection of information, another of its processing and analysis and the third of distribution of information to consumers. As an example, consider the experience of Russia (Фролов и др, 2003; Smirnov et al. 2000).

Let us to consider the experience of Russia (Фролов и др, 2003; Smirnov et al. 2000).

Until the mid 1980's, common hydrometeorological support for the Arctic region in Russia did not exist. During navigational period, Arctic seas have been divided into zones of responsibility between the Headquarters of Marine Operations (HMO) that included scientific and operational teams providing hydrometeorological support of navigation. The main source of information on the state of ice in the Arctic seas and in the lines the Northern Sea Route was visual aerial ice reconnaissance.

As satellite information on the ice cover state in the Arctic seas had started to be used more widely, the development of the unitary system of hydrometeorological support became a matter of time. Such a system was established in the late 1980s. It was mainly based on the "Automated Ice and Informational System for the Arctic" alternatively known as "North" (Бушуев и др., 1977; Фролов и др. 2003).

Infrastructure of the "North" system is presented by territorial hydrometeorological centers (Murmansk, Arkhangelsk, Dikson, Tiksi, Pevek), regional centers for receiving and processing satellite data (Moscow, Yakutsk, Khabarovsk). AARI is the leading center of the "North" system.

Contemporary Russian system of monitoring of the sea ice in the Arctic water areas provides collection, processing and distribution of the ice information to customers practically in the real time and has a unitary center of Ice and Hydrometeorological Information (AARI).

The main source of data on the state of the ice cover in the Arctic and the freezing seas is a satellite remote sensing (information comes from satellites NOAA, Fengyun and EOS (Terra, Aqua), RADARSAT1, Envisat). Surface-mounted receiving complex of AARI in Saint Petersburg is equipped with stations for receiving satellite information (Telonics, USA and ScanEx, UniScan-36, Russia) and provides satellite images in the real time.

Other sources of information about the environmental state in the Arctic are the network of polar hydrometeorological stations, expeditionary research vessels; automatic meteorological buoys deployed on the drifting ice in the Arctic Ocean; domestic and foreign centers of hydrometeorological and ice information;

Center of Ice and Hydrometeorological Information is in charge of:
− Composing review and detailed ice charts;
− Making long-term ice and weather forecasts;
− Making medium- and short-term ice, meteorological and hydrological forecasts;
− Elaborating recommendations on navigation at performance of the marine activities.

Transfer of the real-time information products (charts, forecasts and other information) directly to customers is carried out by means of conventional and satellite communication channels. To transmit information to large ships and icebreakers equipped with INMARSAT communications systems and electronic cartographic navigational and informational systems (ECNIS / ECDIS), special technology developed at AARI is used.

The basic principle of the system is that all information products should be prepared in the unitary center at special "automated work places" (AWP) and then transferred directly to the Captain's workplace (i.e. conning bridge) − to the "End User" AWP. This system represents a technological complex, which is called "Adaptable complex for monitoring and forecasting of the atmosphere and the hydrosphere state for support of marine activities in the Arctic and in the freezing seas of Russia" ("AK-

MON") (Миронов и др., 2009). "AKMON" allows us to adjust the process of monitoring of the environmental state to the specific physical and geographical characteristics of the work area and the specific needs of the costumer. On the bridge, Captain of the vessel can obtain an image illustrating ice conditions combined with a navigation map.

3 PURPOSE OF IM SYSTEM

Experience of comprehensive marine operations in ice demonstrated that informational support with data on the current and foreseeable hydrometeorological and ice conditions is not enough for the efficient and safe functioning of complex technical systems (producing platforms, unloading terminals) and transport systems (transportation of oil products by tankers, the work of the tanker-platform system). Considering the enormous maintenance cost of the facilities and potential environmental consequences (e.g., an accident in the Gulf of Mexico in 2010) as well as the complexity of the environment conditions of the Arctic region, it is necessary to perform a set of mutually dependant informational and logistical activities, that must be based upon the IM system in order to come to an effective management decision.

IM system includes permanent monitoring of the ice conditions, forecasting, assessing the risk of any possible ice conditions, preparing recommendations on the most effective decisions. Final procedure in the sequence of the IM system's steps is the activates aimed at suspension of marine operations (such as full stop of production-unload operations on the platform), re-scheduling (e.g., choice of alternative route for tanker) or the use of various technical means to influence the ice cover for eliminating danger (e.g., towing bergy bit in a distance where it doesn't threat the platform).

"Ice management" system is not an absolutely new invention. Currently, as the world practice and Russian one in particular are concerned, a certain experience has been obtained in navigation in the ice by means of ice-breakers of different class and the use of drilling platforms in the freezing seas. To reduce the risk of the activities and prevent of dangerous impact of ice and icebergs on the vessels, floating and immovable platforms and terminals, it is necessary to manage ice conditions (IM).

Different countries and organizations acting in the Arctic started to implement various IM systems, which at this stage of development were focused on specific local operations and types of activities.

Following examples illustrate the most successful cases in the IM system arrangement:
− Ice management in the region of Great Banks of Newfoundland, where drilling platforms and FPU

function, with use of different methods to change the drift of icebergs (Comprehensive…, 2005).
- Ice management to support high-latitudinal experimental drilling in the drifting ice of the Central Arctic Basin ("ACEX-2004" project) with a mobile drilling platform (Юлин, 2007).
- Ice management to support work in the ice of special unloading equipment for loading oil tankers near the port of Varandei (south-eastern Barents Sea).
- Ice management to support work in the ice of special unloading equipment for loading oil tankers in De-Kastri ("Sakhalin-1" Project) (Herbert, Mironov, 2007).
- Ice management to support loading of oil tankers near the complex "Vityaz" ("Sakhalin-2" Project, phase 1).

4 THE CONCEPT OF THE IM SYSTEM'S STRUCTURE (IM STRUCTURE BY THE EXAMPLE OF THE SHTOKMAN GAS CONDENSATE FIELD)

All of the abovementioned examples of the IM system's arrangement had the same disadvantage as they were designed for the particular marine operation or a specific local area and consequently considered and were limited by the needs of this operation. Nevertheless, widening of activities range in the Arctic imposes the need to develop a universal IM system. In this particular case, universality means the possibility to adjust the system, based on common arrangement principles, to any operation and any region of the Arctic.

It is the main objective of the "Ice Management" system to provide safety of the vessels and offshore facilities whilst functioning in the presence of the sea ice and icebergs in difficult meteorological conditions of freezing seas.

The main common principles of the arranging ice management system are:
- The system should be structured in such a way that to provide high reliability level under any environmental conditions and any conditions of functioning;
- IM system should decrease frequency of interaction of offshore objects with ice formations
- The system should be able to reduce ice loads on the objects when it is impossible to avoid them;
- Activities connected with the ice management should ensure safety of the facility, cut down facility idle time, provide safe and effective disconnection of the facility, accelerate its removal and the safe re-connection.

According to the international standard ISO 19906 (2009), the ice management system should consist of the following subsystems:

1 Subsystem for monitoring and forecasting the ice cover state and distribution of icebergs;
2 Subsystem for evaluation of potentially dangerous ice phenomena and ice formations;
3 Subsystem, using various technical means for effecting the ice cover and icebergs;
4 Subsystem for preparing the facility to a hazardous situation and ensuring its disconnection and removal.

5 STRUCTURAL AND FUNCTIONAL SCHEME OF IM SYSTEM (BY THE EXAMPLE OF SGCF)

The suggested IM system (Shtokman Gas Condensate Field (SGCF) region as an example) should consist of four listed sub-systems, each developed and involved to a different degree of development, taking into account the regional and functional requirements for marine operations. Structural and functional scheme consists of 4 subsystems:
- *Subsystem for monitoring* provides a regional monitoring of icebergs and the ice cover in the Barents Sea and includes modules for collecting and processing of global and regional data, modules for analysis and forecasting, for control of technological processes and transfer of information products.
- *Subsystem for the threat evaluation* provides local monitoring within a specified radius around the producing vessel (platform) and the estimate of probability of dangerous ice phenomena appearance, taking into account the specified radius of risk.
- *Subsystem for influence on ice* lets to develop recommendations for active influence on the ice by means of icebreakers and vessels (tugboats) of various ice class taking into account characteristics of ice regime and helps to make the decision on which technical means to use.
- *Subsystem for preparing* helps to develop recommendations on procedures to begin disconnection and removal of the facility and then, reconnection.

6 ARRANGEMENT SCHEME OF IM SYSTEM (BY THE EXAMPLE OF SGCF)

6.1 *Regional and local monitoring.*

Region of SGCF is located in the margin zone of the Barents Sea, where the sea ice and icebergs aren't recorded every year. If ice enters the field area, average duration of the ice period comprises about 2 months and the maximum – 5.5 months (Наумов и др., 2003). In addition, SGCF area is situated in bergy waters, i.e. there's no sea ice but the probability of encountering icebergs exists.

Taking into account the multiyear and seasonal characteristics of the sea ice and icebergs variability, as well as the high maintenance costs of local monitoring of the ice cover on board the vessel (platform), it seems reasonable to use dual scheme arrangement (fig 1):

1 Regional monitoring of the ice cover and icebergs is carried out on the base of the coastal center of the automated ice and informational system of AARI (Saint-Petersburg).

2 Local monitoring in the Shtokman GCF region is performed from the platform. In the case there's probability of appearance of ice and icebergs in the field region, immediate response group is directed to the platform to carry out additional observations and analyze the potential risk for ice formations to come in the immediate vicinity of the platform.

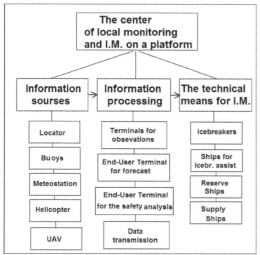

Fig. 1. Arrangement structure of the Ice Management System (fragment)

Regional monitoring of icebergs and the ice cover in the Barents Sea should be carried out throughout the year. Data of the regional monitoring data should be used to make long-term forecast of the ice edge position, and the medium-term forecast of ice distribution and drift of icebergs which later will help to make a decision on the beginning of the group's work on the platform.

In the coastal center of the regional monitoring, module for the collection and processing of information including station receiving satellite images and telecommunication center should function in the 24-hours mode. Module for creating informational products as well as the one for managing technological processes operates in the mode agreed on with the Customer.

Local monitoring is the main source of information of the subsystem for the threat evaluation and is performed by the technical means installed immediately on the platform or near the producing platform.

To achieve objectives of the local monitoring on the platform, and near it, it is necessary to use various equipment: ice radar, helicopter, set of unmanned aerial vehicles (UAVs); camcorders, spotlights, meteorological complex; buoy-markers (for GPS-tracking); Doppler profiler and current meters.

Center of local monitoring should function all year round in the 24-hours mode. On the platform, it is necessary to receive information from the regional center, perform 24 hour ice observations, to analyze general and local information (both current and prognostic), to estimate the possibility of the ice threats. If necessary, helicopter is used to identify potentially dangerous objects.

7 ANALYSIS OF THE ICE THREATS.

Ice conditions can be divided into two groups by the way of manifestation, formation, period of existence and an impact on the technical structure:

1 Dangerous ice phenomena.

2 Dangerous ice formations.

Dangerous ice phenomena (DIP) result from the influence of the atmosphere and the drifting sea ice, usually formed by dynamic factors, appear suddenly, active in the limited area and for a limited period of time. The main way to avoid them is the short-term forecasts with and advance time from several hours to 3 days

Dangerous ice formations (DIF) are solid objects, floating on the surface of the sea area, which physically affects the technical structure of its mass. The main way to avoid their influence is early detection, long- and short-term forecasts of the ice formation's motion and, if necessary, affect them by technical means.

For the SGCF region the following DIP can be distinguished: intense ice drift, compacting, icing (aerial and spray). Dangerous ice formations for the SGCF are, in our opinion, giant, vast, and big floes (breccia) of the first-year ice, any floes and medium floes of the significantly deformed first-year ice, floebergs, icebergs, bergy bits and growlers.

Taking into account possible threats, the time they were discovered, the time various technical means were applied to them, different detection radius and the time required for analysis, coming to decision and implementation of certain actions, table of levels and thresholds of allowable risks can be composed (Table 1). This table presents levels of threat, time needed for necessary measures and recommendations for IM activities.

In order to illustrate the threat level easier, range of colors (from black to green), a widely used in the ice management practice, is associated with description.

Table 1. Evaluation of the level of threat of a dangerous situation and recommendations on the ice management

Threat level	Time required	Colour	IM actions
No threat	Not limited (more than 5 days)	Green Vessel/ Platform works	Regular ice monitoring
Potential threat	From 3 to 5 days	Blue Vessel/ Platform works	Regular ice monitoring
Insignificant threat Distance from the facility is more than 50 km	From 2 to 3 days	Yellow Vessel/ Platform works	Regular ice monitoring Preparation to the possible use of technical means
Real threat Distance from the facility is 10-50 km	From 1 to 2 days	Pink Vessel/ Platform works	Use of all the possible technical means to eliminate the threat Decision to use additional tech. means
High level of threat Distance from the facility is 4-10 km	from $T_{pred} = T_{min} + T_{crit}$ to 1 day	Red Vessel/ Platform works Readiness to disconnect vessel/platform	Use of all the available and additional technical means to eliminate the threat Decision to disconnect the vessel/platform
Critical threat Distance from the facility is less than 4 km	Less than T_{pred}	Black Disconnection of the vessel/ platform	Immediate disconnection of vessel/platform at the same time with active influence on the threat by the all technical means

8 METHODS PERFORMED BY TECHNICAL MEANS AIMED AT DESTRUCTION OF ICE FLOES AND CHANGE OF THE ICEBERGS DRIFT

Protection of the marine facilities against the pressure, piling and the stress caused by drifting ice and icebergs is rather complicated and expensive problem, but technically it can be solved. The task is to destruct large forms and monoliths of ice to the parts of smaller sizes, that freely moves around the structure without creating a critical pressure and stress.

Technological solution of the problem is to breaking down large ice formations (floes, medium floes, small floes) to the smaller part in front of the protected structure. In general, allowable sizes of the ice formations are those with sizes of less than 20 m in

diameter. Gained experience in the ice management shows that there're one or several icebreakers required (depending on ice condition) in order to influence the ice on the ice if you want a few icebreakers, depending on ice conditions (Юлин, 2009).

Various methods to influence icebergs were applied to provide safe production of hydrocarbons in the Great Bank of Newfoundland and in the Labrador Sea. Under the influence of strong Labrador Current, carrying icebergs on to the producing platform, a sufficient security measure is to change the trajectory of the iceberg by a few degrees deflection from its initial courses.

The most commonly used techniques are towing with help of synthetic cable (72% of all of the operations), the deflection by propeller stream (9%), towing by net (7%) and the deflection by water cannons (6%). In the rest cases of deflection of icebergs were performed by less common means (towing by two vessels etc.). (McClintock et al., 2007). Main factors limiting the use of different means to change the iceberg drift are their weight and the wave height. Depending on the wave height, towing of icebergs can be carried out successfully in from 69% to 85% of cases; however, if the wave height is more than 5 meters, the success rate drops to 60%.

Icebergs of almost all the sizes (except for small icebergs and bergy bits) can be towed with efficiency of 76-80%. The average time of influencing the iceberg for a various methods ranges from 4 hours (in case of use propellers) to 8.6 hours (in case of towing by two vessels) (Comprehensive ..., 2005).

From the works (Crocker et al., 1998; Comprehensive..., 2005; McClintock et al., 2007) we can conclude that as a rule for each size gradation of the iceberg, the most effective method is different. Thus, the most effective way to change the course of growlers and bergy bits is the deflection by the propeller stream. Bergy bits and small icebergs can be successfully deflected the by the water cannons. Bergy bits, small icebergs and even medium-size icebergs can be towed by special nets. Method for towing medium and large icebergs, proved its reliability, is the towing by one vessel wit use of synthetic rope (AARI has such an experience, described in Stepanov et al., 2005). Towing large and giant icebergs requires two vessels (icebreakers).

In the area of SGCF it is possible to detect drifting hummocked ice floes, which can be broken into smaller parts not threatening the process of hydrocarbons production by technical means if necessary. If icebergs or bergy bits enter the area, there can be a situation when it's impossible influence the trajectory of their drift effectively and there is a high probability of collision with the facility. In this case, if the criteria are specified, there is a need to implement measures to stop production, disconnect the producing vessel and remove it into a safe distance.

The ultimate measure to eliminate the ice threat, letting to avoid damage of producing vessel, is its disconnection and removal to a safe distance. At the same time, it should be admitted that this measure is highly undesirable due to interruption of the producing cycle (the decrease in the amount of raw) and the subsequent resuming of production (which may require considerable time).

9 THE PROSPECT OF CREATION OF THE UNITARY INFORMATIONAL SPACE AND SECURITY SYSTEMS FOR TRANSPORT OPERATIONS IN THE ARCTIC AND FREEZING SEAS.

Development of the marine and economic activities in the Arctic and other freezing water areas, the complex character of environmental conditions, arising technological and environmental risks strongly require new forms of information service for these comprehensive industrial and transport systems, which include exploration, production, loading and transportation of raw materials.

The most promising way in our point of view is the development and implementation of innovative information and logistics systems, "Ice Management" systems

Russia and other countries, acting in the Arctic, have accumulated wide experience in development and use of the individual elements of these systems.

By now, the concept of IM system was elaborated in Russia for regional conditions of Shtokman GCF. Individual elements of the system already have been successfully working in a number of projects. Analogous systems can be adjusted to other objects of activity in the Arctic region. Use of the IM systems for activities in the Arctic will allow to reduce risks caused by environmental conditions and to make operation of technological systems safer and more effective.

REFERENCE

Бушуев А.В., Волков Н.А., Гудкович З.М., Новиков Ю.Р., Прокофьев В.А. Автоматизированная ледово-информационная система для Арктики (АЛИСА) // Тр. ААНИИ. 1977. Т. 343. С. 29-47.

Миронов Е.У., Ашик И.М., Бресткин С.В., Смирнов В.Г. Адаптируемый комплекс мониторинга и прогнозирования состояния атмосферы и гидросферы // Проблемы Арктики и Антарктики, №3 (83), 2009, с. 88-97.

Наумов А.К., Зубакин Г.К., Гудошников Ю.П., Бузин И.В., Скутин А.А. Льды и айсберги в районе Штокмановского газоконденсатного месторождения // Тр. Межд. Конф. «Освоение шельфа Арктических морей России (RAO-03) – СПб, 16-19 сентября, 2003. – с.337-342.

Фролов И.Е., Данилов А.И., Бресткин С.В., Миронов Е.У. Автоматизированная ледово-информационная система для Арктики и ее использование при освоении углеводородных месторождений на шельфе // Тр. Межд. Конф. «Освоение шельфа Арктических морей России (RAO-03) - СПб, 16-19 сентября 2003, с. 304-307.

Юлин А.В. Основные результаты ледовых наблюдений в высокоширотной арктической экспедиции «АСЕХ-2004» // Проблемы Арктики и Антарктики, № 77, 2007, с. 107-114.

Comprehensive Iceberg Management Database Report 2005 Update, PERD/CHC Report 20-72, Provincial Airlines Environmental Services, Canada, 2005, 182 p.

Crocker, G., Wright, B., Thistle, S. and Bruneau, S. An Assessment of Current Iceberg Management Capabilities. Contract Report for National Research Council Canada, Prepared by C-CORE and B. Wright and Associates Ltd., C-CORE Publications 98-C26, p 105, 1998.

Herbert J.C., Mironov Ye.U. Marine transportation in ice //Тр. Межд. Конф. «Освоение шельфа Арктических морей России (RAO-07), 11-13 сентября 2007, СПб, (Электронная версия на CD No. 213).

McClintock J., R. McKenna and C. Woodworth-Lynas. Grand Banks Iceberg Management. PERD/CHC Report 20-84, 2007, 92 p.

Smirnov V., Frolov I., Grishenko V. and Mironov E. Russian ice information services in the future // Proc. of the Northern Sea Route User Conference, Oslo, 18–20 November 1999, Kluwer Academic Publishers, Dordrecht, 2000, p. 169-176.

Stepanov I., Gudoshnikov Yu., Iltchuk A. Iceberg Towing Experiment in the Barents Sea // Proc. of the 18th Int. Conf on Port and Ocean Engineering under Arctic Conditions (POAC-2005), Potsdam, NY, USA, June 26-30, 2005, vol. 2, pp. 585-594.

Ship Construction

10. Investigations of Marine Safety Improvements by Structural Health Monitoring Systems

L. Murawski, S. Opoka, K. Majewska, M. Mieloszyk & W. Ostachowicz*
Institute of Fluid-Flow Machinery PASci, Gdansk, Poland

A. Weintrit
*Gdynia Maritime University, Faculty of Navigation, Gdynia, Poland

ABSTRACT: The paper presents a first approach of the structural health monitoring (SHM) system, dedicated to marine structures. The considered system is based on the fibre optic (FO) technique with Fibre Bragg Grating (FBG) sensors. The aim of this research is recognition of possible practical applications of the FO techniques in selected elements of marine structures. SHM and damage detection techniques have a great importance (economical, human safety and environment protection) in the wide range of marine structures, especially for ships and offshore platforms. The investigations reported in this paper have shown major potential of FBG sensors. They are suitable for strain–stress field and load monitoring of the wide range real structures used in different conditions. The FO sensors technology appears as very attractive in many practical applications of future SHM systems.

1 INTRODUCTION

In this paper authors presents examples of FBG (*Fibre Bragg Grating*) sensors applications in the SHM (*Structural Health Monitoring*) technology for marine structures. The idea of the SHM is to build a system that contains many sensors and is able to evaluate the structure condition. One of the most promising sensors are those based on FO technology, especially FBG sensors. The first structure on which the FBG sensors were mounted assigned to measure strain and temperature was Beddington Trail Bridge – Calgary (Canada) described by Measures et al. 1993. In the following sections, the authors describe applications of this enabling technology to SHM of marine structures like Horyzont II ship and offshore structure model.

Because of their advantages like, high corrosion resistance, immune to electromagnetic interference and multiplexing capabilities, FBG sensors can work properly in the harsh marine environment for many years (up to 20) (Lee 2003) and are very interesting for SHM systems. Application of SHM techniques allows one to increase both human and environmental safety in marine industry with simultaneous reduction of maintenance costs.

1.1 *FBG sensors*

A Bragg grating is a permanent periodic modulation of the refractive index in the core of a single mode optical fibre by exposing the core of the optical fibre

to an interference pattern of intense UV-laser light (Bass et al. 2001). The length of the FBG sensor is in the range 1-25 mm and depends on the application. This periodic perturbation in the core index of refraction allows coherent scattering to occur for narrow wavelength band of the incident light travelling within the fibre core. A strong narrowband back reflection of light is generated, centred around the maximum reflecting wavelength value λ_B, when the resonance condition is satisfied:

$$\lambda_B = 2n_{eff}\,\Lambda \qquad (1)$$

where n_{eff} is the effective refractive index and Λ is the periodicity of the perturbation (Udd 2006). The schema of FBG sensor is shown in Figure 1. The λ_B is dependent upon the geometrical and physical properties of both the grating and optical fibre. The key point of these FBG sensors is their wavelength-encoded nature, which is an absolute parameter providing reproducible measurements (Udd 2002).

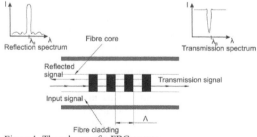

Figure 1. The schema of a FBG sensor

The FBG sensors are sensitive to both strain and temperature. Because of the cross sensitivity on those two parameters in the case of strain measurement additional sensors dedicated to temperature sensing are needed. The changes in those two parameters are linearly proportional to changes in measured wavelength (Udd 2006).

Most of the conventional sensors used in SHM applications are based on transmission of electric signals. These sensors are usually not small. They are local sensors and cannot be easily multiplexed and embedded in a structure. There is also a problem with protection against the corrosion processes in marine environments. Conventional strain gauges mounted on an offshore platform CB271 were completely unusable after one year in Bohai Sea, while the FBG sensors mounted close to them work properly (Ren et al. 2006). In some cases, the signals from electric strain gauges could not be discriminated from noise because of electrical or magnetic interference. On the contrary to strain gauges the FBG sensors have small size and weight, multiplexing capabilities and are immune to electromagnetic field and have high corrosion resistance. They can also be mounted onto the structure (Ren et al. 2006) or even embedded (Wang et al. 2001) into material of an element during its manufacturing. Those advantages make them to be a very interesting tool in SHM systems mounted on marine structures especially for those made from composite materials.

2 MARINE STRUCTURES

Marine structures like marine vessels and offshore structures surrounded by a harsh marine environment are exposed to long-term cyclic loadings comes from continuously acting sea waves and short-term extreme loads such as severe storms, seaquakes or collisions. The marine environment (sea water) results in fast corrosion, erosion and scour processes. Those phenomena increase the size of existing damage and also initiate its growth. Any damage of marine structure can results in ecologic catastrophe because of the oil which is extracted by offshore platforms or fuel in marine vessels. Another important point for using SHM systems is increasing the safety level for people working on marine structures.

Nowadays the newly designed structures are designed using FEM (*Finite Element Method*). The next step are the sea trials for a prototype. The designers are interested with strain and temperature distribution over the structure. At the designer level those information can be then utilized for optimization of the structure like it was in the case of Fast Patrol Boat HnoMS Skjold (Wang et al. 2001).

The SHM systems can be also used on marine structures during they exploitation. Nowadays developed SHM systems for marine vessels are especially designed for ships hulls monitoring. Such systems based on FBG sensors exist for example on Fast Patrol Boat HnoMS Skjold (Wang et al. 2001), minecountermeasure vessel HnoMS Otra (Torkildsen 2005).

2.1 Marine vessels

SHM systems based on FBG sensors can by performed for both prototypes (Wang et al. 2001) and serial production vessels (Torkildsen et al. 2005). Those systems are especially used for ship hull monitoring. The FFI (*Forsvarets forskningsinstitutt, Norwegian Defence Research Establishment*) has been involved in the development of a SHM system based on FBG sensors since 1995 (Torkildsen et al. 2005). In 1999 the first system CHESS (*Composite Hull Embedded Sensor System*) was installed on a prototype HnoMS Skjold of the Skjold class Fast Patrol Boat in the purpose of SHM of composite hull and collecting information about its real loading achieving under normal work (Wang et al. 2001). Nowadays there is the extended SHM system for hull monitoring developed by FFI that contains different kind of sensors for measuring air pressure, profile of meeting wave, weight distribution and GPS for determination the speed and position of the ship (Figure 2) (Torkildsen et al. 2005).

The idea use FBG sensors is to monitor static and dynamic strains at critical positions of the hull. The sensors should be located for measurements of global moments and forces acting on the hull (Torkildsen et al. 2005).

The purpose of using SHM system for ship hull monitoring is verification of model results, condition-based maintenance to reduce cost and non-operative periods, damage detection and evaluation of residual hull strength. An extended SHM system will also include monitoring and recording of the ship motion, operating parameters and sea waves parameters (Torkildsen et al. 2005). The results are presented to the crew on the bridge, in the machine control room, and operational rooms (Torkildsen et al. 2005). Alarms are activated when the hull loads approach the ship design limits (Torkildsen et al. 2005).

The advantages of systems based on FBG sensors are especially visible for implementation of new materials and construction designs. Because of the advantages of composite materials like, a high strength-to-weight ratio, good impact properties and low infrared, magnetic and radar cross-sectional signatures, design versatility (Mouitz et al. 2001) nowadays there is a wide range of naval structures being developed using FRP (*fibre reinforced polymer*) (Rao 1999). This development is driven by the need to enhance the operational performance (e.g. increased range, stealth, stability, payload) but at the same time reduce the ownership cost (e.g. reduced maintenance, fuel consumption costs) marine vessels (Mouitz et al. 2001).

Figure 2. Overview of environmental and operational sensors on HnmMS Otra (based on Torkildsen et al. 2005)

Composite materials have been used for the construction of naval patrol boats since the early 1960s. The first all-GFRP patrol boats were built for the US Navy and were used during the Vietnam War. Nowadays the largest all-composite naval patrol boat is Norwegian HnoMS Skjold of the Skjold class Fast Patrol Boat (Mouitz et al. 2001). The boat was built in 1999 (Wang et al. 2001). HnoMS Skiold is a twin-hull SES (*Surface Effect Ship*) made of fibre reinforced polymer composites sandwich panels (Wang et al. 2001, Mouitz et al. 2001). The boat is 47 m long, 13.5 m wide and its weight is about 270 tons. A SES is a principle based on a catamaran hull where lift fans blow air into an air cushion trapped between the hulls. HnoMS Skjold was designed for high speed – for example 45 knots at sea state 3 (Wang et al. 2001).

On a prototype HnoMS Skjold a system CHESS was installed in the purpose of SHM of composite hull of the boat and collecting information about its real loading achieving under normal work. The project was a join between the US Naval Research Laboratory, Washington, DC and the Norwegian Defence Research establishment. The most important application of the system was verification of assumed balances the weight optimization and structural strength requirements of the composite hull (Wang et al. 2001).

The first step during SHM system design was created a FEM model allowing for define strain/stress field in the hull loaded by sea waves during different sea states. Calculated strain/stress fields were used for determine a localisation of FBG sensors embedded into composite hull (Wang et al. 2001). The chosen locations (Figure 3) allowed for analysis the most significant wave loads like: vertical bending (hogging/sagging), horizontal bending, torsion (twisting moments), vertical shear force and longitudinal compression force (Wang et al. 2001).

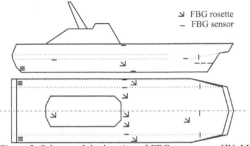

Figure 3. Schema of the location of FBG sensors on HNoMS Skjold (based on Wang et al. 2001))

The systematic sea-keeping tests using the CHESS system on HnoMS Skjold were performed in May to June. Registered dates were collected during sailing thought the sea and routine operations with the vessel. There were mounted few conventional strain gauge in the purpose of verification of strain measured by an array of 56 FBG sensors. The agreement between those two methods was good, but data from strain gauge were disturbed by electromagnetic field influence. Basing on the data from the SHM system important changes in HnoMS Skjold boat design were performed (Wang et al. 2001).

2.2 *Offshore platforms*

Offshore structures surrounded by a harsh marine environment are exposed to long-term levels higher of cyclic loadings comes from continuously acting sea waves, accumulating of floating ice shocks, and short-term extreme loads such as severe storms, seaquakes and accidental collisions. Additionally they are exposed to corrosion, erosion and scour. Those phenomena increase size of existing damages (Ren et al. 2001, Sun et al. 2007).

Because of existence of thermal errors, large zero drifts, non repeatable readings, difficult signal conditioning and high susceptibility to moisture and corrosion influence of sea (cauterization of sea), electrical sensors are restricted in the offshore platform application. Those disadvantages do not occur for FBG sensors (Ren et al. 2001, Sun et al. 2007).

Bohai Ocean Oil Field is one of main ocean oil fields in China. There is very heavy ice force in winter which become the main environmental force of offshore platforms. In 1969 and 1977 two platforms collapsed by heavy ice force action. Since 1980's, the ice conditions, ice pressure on and response of platforms in Bohai ocean have being monitored under support of China Ocean Oil Company (Ou et al. 2004).

One of the SHM systems was installed on CB32A steel jacket platform in Bohai under the project supported by the National Hi-tech Research and Development Program of China. The Platform with jacket height 24.7m will be built in 2003 and located in water depth 18.2m. The system includes 259 FBG sensors, 178 polivinylidene fluoride sensors, 56 fatigue life meters, 16 acceleration sensors and a set of environmental condition monitoring system (Ou et al. 2004).

Sun at al. 2007 built a model of an offshore platform CB32A in the scale 1:14. The model consisted of steel pipes is 2.69 m tall, 1.55 m long and 1.5 m wide. There are (Figure 4) 7 FBG strain sensors, two accelerometers and one temperature compensation FBG sensor installed on the structure. A recording, storage and interrogation system was put in the office occurred about 100 m from the model. This was in the purpose of investigating the ability of making measurements using a long distance system (Sun et al. 2007).

One of the investigated dynamic loading was Tianjin wave which is a kind of seismic wave measured for the first time in the base of Tianjin hospital in Tangshan in China and named from the place of the first appearance. Tendency of responses from two types of sensors (the FBG strain sensors and electrical strain gauge) were the same and maximal values of measured parameter were close to each other. The level of a disturbances was about 10 με and only 1 με for strain gauge and FBG sensors, respectively. In the compare with strain gauges, FBG sensors showed its particular feature which is insensitivity on influence of electromagnetic field and high ability of dynamic strain measurements in systems with low vibration frequencies (Sun et al. 2007).

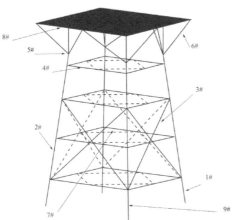

Figure 4. Location of FBG sensors on offshore platform model: 1# – 7# – strain sensors, 8#, 9# – accelerometers, 10# – temperature compensation sensor (based on Sun et al. 2007)

FBG strain and temperature sensors were implemented into SHM system on steel CB271 offshore platform located in the Bohai Sea, of East China. A FBG sensors rosette was located on the bottom side of a middle support. Temperature compensation FBG sensor was located close to it. The FBG sensors were covered by epoxy layer in the purpose of protect under destructive influence of environment. This system correctly monitor on line responses of the structure (its strain) under loading of sea waves and waves generated by several hundred-tone ships sailing close to the platform (Ren et al. 2006).

The state of the FBG sensors array was checked one year later. They were working as good as in the moment they had been mounted. No significantly decrease of the sensitivity was observed. However electrical strain gauges located close to them were damaged because of corrosion influence of the sea water. This allowed the FBG sensors to show its advantages allowed them to being a part of SHM system permanently installed on offshore structure (Ren et al. 2006).

One of the most dangerous accidents being able to damage the offshore structure and installation mounted on it is a collision with a vessel. Because of this a registration of changes of strain during such accidents is very important. Such accident was happened and was observed in sensors measurements on 20 July 2004. There was no dangerous of damage occurrence because the value of the registered strain induced by the ship impact was in the linear-elasticity range of the material. Equally important like a ship collision is loading from ocean waves. The level of the strain introduce from strong ocean wave impaction of the ocean can be closed to those measured during the ship collision (Ren et al. 2006).

3 DAMAGE DETECTION EXPERIMENT BASED ON MEASURED STRAINS

Certain number of the offshore production platforms exceeded their working life assumed by design engineers. Growing possibility of the catastrophic failure should be prevailing for production stop and removal of such objects. On the other hand prolonging the production beyond working life makes extra financial profits. In such situation the optimal solution is to install a SHM system on such structures which early warns about a structural problem.

Before starting field investigations there is a need for checking SHM systems on some laboratory-size models. In the first step a space frame leg model of typical offshore jack-up rig was constructed and tested, see Figure 5. The model consists of 3 main chords which are connected by horizontal and diagonal brace elements.

The braces of the structure were covered by 16 FBG strain sensors located at bay level 2 and 7 counting from the bottom. Additionally 6 bare FBG sensors were glued on each chord. Also 2 FBG acceleration sensors were mounted in the upper part of the structure.

Figure 5. The leg model of jack-up rig with 1500 kg top mass.

The behaviour of the leg model was analyzed under different conditions. Damage was located at the bottom part of the model and it was modelled as chord's yielding, partial and also entire chord cutting (with open and closed gap) and cutting of the bottom K-brace of the leg model. Concurrently different loading scenarios was taken into account: hammer impacts to impart free vibrations of the model and 10kN horizontal shaker to induce forced vibrations. Vibration experiments were done for unloaded leg model, leg model with added 1500 kg top mass and the leg model loaded by vertical load up to 500 kN from hydraulic actuator situated on the floor.

The main aim of the investigations was the identification of damage influence on readings from sensors glued to different members of the leg model.

In this paper the authors consider here only the static loads and show the results of 2 selected sensors for one damage scenario i.e. chord's yielding at bay level 1, see Figure 6.

The selected sensors are those glued on chord A, one at bay level 2 (denoted as the closest to the damage) and second at bay level 7 (denoted as the farthest from the damage). The corresponding time recordings are shown in Figure 7 and Figure 8. Please note that chord A is in compression state as a result of added top mass and applied vertical load. Readings show that in region close to (far from) the dam-

age increase in tension (compression) strains occurs. In both cases yielding of the material forces significant strain changes. The same behaviour is observed on remaining sensors glued on the chords. Only the difference between strain levels and the direction of change is dependent on sensor choice. The strain readings were recorded before and after occurrence of yielding. Therefore, plots in transient region are not continuous because the transient state can be different than these simple jumps.

Figure 6. Bottom part of chord A after yielding.

The most promising is the fact, that local yielding can be detected by sensors situated relatively far from the damage.

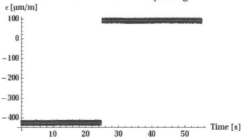
Figure 7. Strain level increase after local yielding.

FBG sensor on chord A farthest from yielding area: strain levels before and after yielding

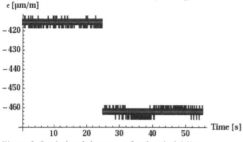

Figure 8. Strain level decrease after local yielding.

4 SCHEME OF SHM SYSTEM ON REAL MARINE STRUCTURE

The authors present a first approach of the SHM system, dedicated to marine structures. The considered system is based on the fibre optic technique, especially on FBG sensors. The aim of this research is recognition of possible practical applications of the fibre optic techniques in selected elements of marine structures. The authors have been performing initial investigations on Horyzont II ship. The FBG based system will be compared with classical measurement techniques, e.g. piezoelectric accelerometers. The environmental loading conditions will be also monitored.

Designed SHM system is located on a mast of Horyzont II ship (see Figure 9). FBG sensors are placed in the bottom area of the mast (see Figure 10) where predicted strain-stress level is the highest. The authors finally planned to use 5 FBG sensors for strain measurements and one FBG sensor for temperature compensation. Three of FBG strain sensors are arranged for a strain rosette. So, strain-stress absolute values with main axis directions can be determined with the rosette. One FBG sensor is located one meter higher than the rosette, just for longitudinal stress distribution determination. Another FBG is placed on the same highest as previous one, but for stress level determination in lateral direction. Three seismic, piezoelectric accelerometers (ACC) are placed on the navigation deck, close to the mast foundation. Seismic ACC can record low frequency movements of the ship in the all directions. Accelerometers give us information about environmental loading, mainly about sea waves excitations. Other environmental loading, like wind force, temperature, atmospheric precipitation, will be registered with other collaborated meteorology system.

The authors worked out methodology of the first step of the measurements. Determination of the dynamic characteristics of the mast is the main target of those investigations. First of all, zero signals will be recorded before and after ship's voyage (in the harbour). Tests on the open sea will be performed in two variants: dynamic (100 Hz scan frequency) and quasi-static (1 Hz scan frequency). The measurements are planned during heavy weaver in Gdansk Gulf area as well as in the open Baltic Sea.

Figure 9. Location of the SHM system based on FBG sensors

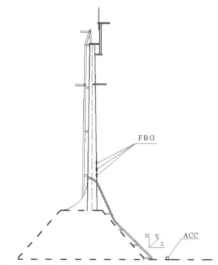

Figure 10. Sensors displacement on the mast

5 CONCLUSION

Marine structures are working in corrosion environment (sea water). Application of SHM techniques allows one to increase both human and environmental safety in marine industry with simultaneous reduction of costs. One of the problems which must to be solved during designing of such system is to find sensors appropriate sensors which could work properly through whole life time of the structure. FBG sensors advantages in compare with conventional strain gauges like immunity to electromagnetic field interference, high corrosion resistance and multiplexing capabilities make them promising tool

88

for SHM technologies implementation to marine structures, like marine vessels and offshore platforms. In the presented examples FBG strain sensors are used as a part of complex SHM systems implementation on a fast patrol boat HnoMs Skjold (Wang et al. 2001), minecountermeasure vessel HnoMs Otra (Torkildsen et al. 2005) of Norwegian Navy. Another marine structures are offshore platforms. Ren et al. 2006 showed that only FO sensors mounted on offshore platform CB271 on Bohai Sea survived a year of installation of the platform. Strain gauges in contrast were completely destroyed by corrosion processes and were unusable for measurements (Ren et al. 2006).

Searching for new constructing solutions and new materials application in combine with high safety and ecologic requirements will result in implementation of SHM systems based on FBG sensors to many marine structures not only the new designed ones.

FO sensors will be a critical technology in many aspects of future SHM systems. FBG sensors are suitable for strain-stress field and load monitoring of wide range of real-world structures under different conditions. The results obtained with the FBG sensors show good agreement with the electric strain gage. Additionally, the FO sensor network has several advantages: it does not suffer from zero drift, it is self-calibrating, it has low mass, it is immune to electromagnetic interference and it has high multiplexing capability. FBG sensors are more reliable in determination of frequency spectrum of the signal then classical electrical strain gauge. FBG sensors are better suited for long-term monitoring systems.

The investigations reported in this paper have shown big potential of FBG sensors for SHM systems dedicated for such difficult structure as marine ships and offshore platform. Structural damage can be detected on the base of: strain-stress field dynamic characteristics (analytical-empirical self learning system), loads level and counter identification for structure uncertainty determination (on the base fatigue), nonlinearities on the base of known load level, changes of mode shapes and frequencies and changes of structural damping characteristics.

The FO sensors technology appears as very attractive in many practical applications of future SHM systems.

ACKNOWLEDGEMENTS

This research was partially supported by Monitoring of Technical State of Construction and Evaluation of Its Life-span (MONIT in polish) project which was co-financed by the European Regional Development Fund under the Innovative Economy Operational Programme.

REFERENCES

Lee, B. 2003 Review of the present status of optical fiber sensors, Opt. Fiber Technol. 9, 57–79.

Rao, Y.J. 1999 Recent progress in applications of in–fibre Bragg grating sensors, Opt. Lasers Eng. 31, 297–324.

Wang, G., Pran, K., Sagvolden, G., Havsgard, G.B., Jensen, A.E., Johnson, G.A., Vohra, S.T. 2001 Ship hull structure monitoring using fibre optic sensors, Smart Mater. Struct. 10, 472–478.

Ren, L., Li, H.N., Zhou, J, Li, D.S., Sun, L. 2006 Health monitoring system for offshore platform with fiber Bragg grating sensors, Opt. Eng. 45, 352–361.

Sun, L., Li, H.N., Ren, L., Jin, Q. 2007 Dynamic response measurement of offshore platform model by FBG sensors, Sens. Actuators, A 136, 572–579.

Udd, E. 2002 Fiber optics, theory and applications. In: Schwartz M, editor. Encyclopedia of smart materials, vol. 1. John Wiley & Sons, Inc., New York.

Udd E. 2006 Fiber Optic Sensors: An Introduction for Engineers and Scientists. John Wiley & Sons, Canada.

Measures, R.M., Alavie, A.T., Maaskant, R., Ohn, M.M. Karr, S.E. Huang, S.H. Glennie, D.J., Wade C., Guha-Thakurta, A., Tadros, G., Rizkalla, S. 1993 Multiplexed Bragg grating laser sensors for civil engineering, Proc. SPIE Conf. on FOS, Boston, 21-29.

Bass, M., Enoch, J.M., Van Stryland, E.W., Wolfe, W.L. 2001 Handbook of Optics, Volume IV, Fiber Optics And Nonlinear Optics, Second Edition, McGraw-Hill, New York.

Mouitz, A.P., Gellert, E., Burchill P., Challis, K. 2001 Review of advanced composite structures for naval ships and submarines, Compos. Struct. 53, 21–41.

Torkildsen, H.E., Grøvlen, A., Skaugen, A., Wang, G., Jensen, A.E., Pran, K., Sagvolden, G. 2005 Development and Applications of Full-Scale Ship Hull Health Monitoring Systems for the Royal Norwegian Navy, Proc. Recent Developments in Non-Intrusive Measurement Technology for Military Application on Model- and Full-Scale Vehicles, Neuilly-sur-Seine, 22-1 – 22-14.

Ou, J., Li, H. 2004 Recent Advances of Structural Health Monitoring in Mainland China, Proc. ANCER, Hawaii, 152-165.

11. Ultrasonic Sampling Phased Array Testing as a Replacement for X-ray Testing of Weld Joints in Ship Construction

A. Bulavinov, R. Pinchuk, S. Pudovikov & C. Boller
Fraunhofer IZFP, Saarbrücken, Germany

ABSTRACT: According to European Standard EN 1712 ultrasonic testing of thin-walled welded joints is mandatory for wall thicknesses of more than 8 mm only. Any thinner components do have to undergo X-ray inspection.

Besides various advantages of X-ray testing viz. high sensitivity to smallest inclusions, high acceptance in the ship building sector and "automatic" documentation of inspection results, there are also several deficiencies (like radiation protection issues, inspection time expenditure etc.) creating a reasonable request for more cost-effective alternatives.

Sampling Phased Array technology introduced by Fraunhofer-IZFP provides significant improvement of flaw detectability also in thin-walled welded joints due to its tomographic approach in processing of signals obtained by ultrasonic phased array transducers. It allows high-quality imaging of welded joints and detection of relevant material flaws. Real-time ultrasonic imaging with tomographic quality offers a great alternative to X-ray testing with respect to inspection speed and modern documentation of inspection results.

The basic principles of Sampling Phased Array are presented in the paper and several application results obtained on welded joints of marine objects are presented.

1 INTRODUCTION / TASK DEFINITION

In ship structural design and assembly standard non-destructive testing methods include visual, eddy-current, liquid penetrant, ultrasonic and x-ray testing as well as leakage tests. Prevalent techniques for welded joint inspection are x-ray and ultrasonic testing as the most cost-effective and efficient in respect for flaw detection.

Non-destructive testing operations are performed in the scope of quality assurance arrangements according to classification instructions, specifications of ship design and regulations of the manufacturing process [1]. Thus a written procedure for welded joint testing has to be established for any new ship design where the inspection areas have to be defined. Through classification instructions the minimum requirements to be fulfilled for non-destructive testing are defined. Those have to be implemented in the written procedures for particular construction elements.

The non-destructive testing of welded joints is described in the currently generally admitted European codes and standards DIN EN 12062, DIN EN 25817, DIN ISO 5817, ISO 6520.

Since many years x-ray testing is considered as a proven technique for inspection of welded joints in ship construction. Generally speaking its advantages and disadvantages can be seen as follows [2]:

Table 1: Advantages and disadvantages of x-ray testing

Advantages	Disadvantages
Sensitive to both surface and volume flaws	Limitations for thick-walled components
Direct documentation of the inspection results by film	Inspection sensitivity is related to the wall thickness
Flaw size and shape can be directly seen and evaluated	Crack orientation must be known for optimal flaw detection
No direct access to the component is required	Defect height normally can't be defined
	Time consuming technique with significant equipment expenses
	Radiation hazard

Though, this method requires significant operating effort for adherence of radiation protection a spatial separation of inspection area is required. Hence no further work can be simultaneously conducted in the neighborhood. This can lead to significant decrease in manufacturing productivity.

Ultrasonic testing of welded joints is a significant alternative to x-ray testing that can be applied, which in general has the following advantages and disadvantages [2].

Table 2: Advantages and disadvantages of ultrasonic testing

Advantages	Disadvantages
Testing of thick-walled components is possible without limitations	Acoustic coupling (surface contact) is required. Limitations due to surface roughness are possible
Evaluation of flaw size, type, orientation can be obtained.	
Fast, cost-effective testing with immediate conclusion about indication	High requirements on inspection staff due to rather complex calibration of UT instrument
Automated or half-automated inspection and evaluation can be implemented	Limitations on flaw detectability due to suboptimal insonification position or flaw orientation
NEW!: Imaging techniques like phased array allow documentation and quantitative evaluation of inspection results	

For being able to replace x-ray by ultrasonic testing the following tasks must be solved:
- Equal or better flaw detectability of relevant flaws compared to x-ray
- Fast representation and evaluation of inspection results
- Cost-efficient implementation of inspection system and inspection procedure
- Mobile inspection system for in-situ applications

2 SAMPLING PHASED ARRAY TECHNIQUE

The novel ultrasonic inspection technique rapidly coming into industrial application is phased array [3, 4]. Phased array testing offers significant advantages for ultrasonic testing (UT) of welded joints due to its extended information content provided by beam steering capability. Hence the combination of mechanical scanning and electronic beam steering increases flaw detectability, since it is being insonified from various angles of incidence.

Phased array techniques may also have their limitations in certain applications with respect to spatial resolution in the far field of phased array transducers or inspection speed while beam steering over a big angle range and finite signal-to-noise ratio of the system can be seen as an advantage.

Sampling Phased Array (SPA) technology developed by Fraunhofer IZFP is a next step in Phased Array technology. On the one side the technique is capable of fast synthesis of phased array ultrasonic signals for arbitrary angles of incidence with focusing in all the depths within the probe near field. On the other hand the back projection and overlapping in the volume the elementary wavelets obtained by SPA according to synthetic aperture focusing technique (SAFT) principles offers the best possible image reconstruction quality.

The SPA technique offers the following practical advantages:
1 Ultra-fast virtual beam sweep for arbitrary angle range
2 Improved sensitivity and resolution in the near field of the transducer
3 Fast 2D / 3D imaging

Unlike conventional Phased Array technique insonifying the inspection volume by directed sound fields under different angles of incidence, Sampling Phased Array performs data acquisition by exciting cylindrical or spherical waves that propagate in all directions. This can be implemented by firing single array elements or applying defocusing delay laws (Figure 1, left). Hence a very wide angle range can be covered after a single shot.

The ultrasonic signals acquired and saved in each probe position for every single array element serve as an input data for image reconstruction. The reconstruction occurred according to the SAFT algorithm [5]. Since the sound field of array elements is very divergent, every time signal (A-scan) received contains overlapped echo-signals from available reflectors in different volume positions. The reconstructed image in one position of linear array visualizes a cut plane perpendicular to insonification surface, the so called sector-scan. For every point within this plane the propagation times from the transmitting elements and back to the receiving elements are calculated. The amplitude values from all A-scans with matching propagation times are added up in each image point [6].

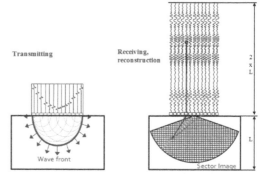

Figure 1: Defocused transmission and sector image reconstruction by SPA

Thus all angles of incidence and focal depths within the near field of the transducer can be realized even after one single transmitting/receiving act. Since the sound beam steering at each volume point, i.e. for all angles of incidence and focal depth, is performed not physically but virtually through the computer, a significant increase in inspection speed can be achieved by implementation of the SPA principle [7]. Furthermore the synthetic focusing in the near field of the UT transducer by the SAFT principle improves sensitivity and resolution (Figure 2).

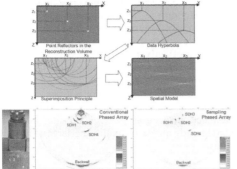

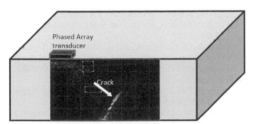

Figure 2: Principle of image reconstruction by SPA

Thus for weld inspection the material flaws can be represented in tomographic quality that allows their exact sizing (Figure 3).

Figure 3: Tomographic Image of an inclined lying crack

3 INSPECTION SYSTEMS FOR INDUSTRIAL APPLICATIONS

Modern instrument engineering, e.g. latest signal processors and computers, offer sufficient computation power for performing SPA image reconstruction and processing, that outmatches conventional phased array systems in speed and quality. Versatile reconstruction techniques [8] can be implemented in a portable manual flaw detector (Figure 4).

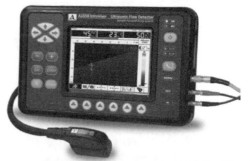

Figure 4: Manual ultrasonic tomograph A1550 IntroVisor by ACSYS

One of the main advantages of ultrasonic testing is its ability to be implemented in automated or semi-automated way, providing fast and cost-effective inspection solutions for industrial applications.

4 ULTRASONIC IMAGING SYSTEMS AS REPLACEMENT FOR X-RAY IMAGING

While position related data acquisition provided by a manipulator or encoder wheel the ultrasonic image can be reconstructed in such a way that it represents the inspected area similar to the X-ray film (Figure 6). The Sampling Phased Array technique with its improved image processing capabilities, e.g. eRDM technique [9], can provide especially sharp and high-contrast images for detecting relevant welding defects. The evaluation of inspection results can be performed based on equivalent flaw size, e.g. by calibrating on artificial defects like notch or side drilled hole or by use of novel image processing algorithms for fast quantitative flaw sizing.

Especially for thin-walled welded joints the Sampling Phased Array technique offers specific advantages due to improved sensitivity and resolution in the near field of array transducer.

SPA System for circumferential weld seams   SPA System for longitudinal weld seams

Figure 5: Semi-automated SPA systems with 2D and 3D imaging capabilities

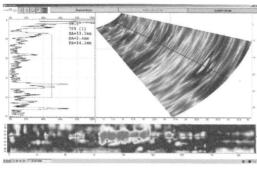

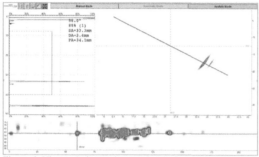

Figure 6: Ultrasonic inspection results on the weld seem with a wall thickness of 6 mm with an elongated cavity in conventional and Sampling Phased Array mode

The current state of standardization of ultrasonic Phased Array testing in Europe is significantly behind schedule when compared to state of the art technology. Codes required like ISO DIS 13588 are in preparation phase.

Fraunhofer IZFP has gathered positive experience in introducing novel phased array techniques to the industrial market as a replacement for X-ray testing of heat exchanger pipes in power plants [9]. The technique was especially developed for testing thin-walled pipes. The ultrasonic testing procedure applied is in accordance to a TÜV Süd specification. Despite novelty of the testing method it could be shown that the ultrasonic imaging provides equivalent performance and reliability like established testing procedures.

5 CONCLUSION

The Sampling Phased Array technique is an enhancement to conventional phased array which does processing of the single shot data sampling taken off-line but in principally real time. Enhanced sensing and data sampling rates allow large data samples to be taken which again lead to 3D images to be generated at comparatively high resolution. Enhanced introduction of SPA into areas where X-ray testing is currently dominating will lead to:

- Comparable or better flaw detectability
- Higher cost-effectiveness
- Prompt evaluation of inspection results
- No radiation protection
- Mobile and stationary inspection set-ups

This may be achieved through handheld as well as automated inspection systems on site, being a considerable advantage in qualifying large ship hull structures.

REFERENCES

[1] U. Cohrs, Anwendung der ZfP im industriellen maritimen Schiffbau, DGZfP-Jahrestagung 2005, 2.-4. Mai, Rostock, DGZfP-Berichtsband 94-CD
[2] V. Wesling, R. Reiter, Zerstörungsfreie Schweißnahtprüfung, Skript zur Vorlesung, Technische Universität Clausthal, Institut für Schweißtechnik und trennende Verfahren
[3] H. Wüstenberg, G. Schenk, Entwicklungen und Trends bei der Anwendung von steuerbaren Schallfeldern in der ZfP mit Ultraschall, Mainz, DGZfP-Jahrestagung 2003, DGZfP-Berichtsband 83-CD
[4] Advances in Phased Array Ultrasonic Technology Applications, by Olympus NDT, January 2007
[5] V. Schmitz, W.Müller, G.Schäfer: Synthetic Aperture Focussing Technique: state of the art. Acoustical Imaging, Vol.19, New York, 1992, pp. 545-551
[6] A. Bulavinov, R. Pinchuk, S. Pudovikov, K. M. Reddy, F. Walte: Industrial Application of Real-Time 3D Imaging by Sampling Phased Array, In: European Conference for Nondestructive Testing, Moscow, June 2010
[7] A. Bulavinov: Der getaktete Gruppenstrahler. Universität des Saarlandes, Saarbrücken, 2005 Dissertation
[8] Samokrutov A.A., Shevaldykin V.G. Ultrasonic Tomography of Metals Using the. Sampling Focus Method. 10th ECNDT, Moscow. June 7 – 11, 2010
[9] R. Pinchuk, A. Bulavinov: Verfahren zur empfindlichen Ultraschallprüfung an rohgeschmiedeten Oberflächen In: Deutsche Gesellschaft für Zerstörungsfreie Prüfung e.V. - DGZfP-, Berlin: ZfP in Forschung, Entwicklung und Anwendung. DGZfP-Jahrestagung 2010. CD-ROM : Erfurt, 10.-12. Mai 2010
[10] R. Weiß, T. Hauke, R. Birringer, S. Caspary, Qualifizierung der Phased-Array-Prüfung als Ersatz für die Durchstrahlungsprüfung, In: Deutsche Gesellschaft für Zerstörungsfreie Prüfung e.V. -DGZfP-, Berlin: ZfP in Forschung, Entwicklung und Anwendung. DGZfP-Jahrestagung 2010. CD-ROM : Erfurt, 10.-12. Mai 2010

12. Conditions of Carrying Out and Verification of Diagnostic Evaluation in a Vessel

A. Charchalis

Gdynia Maritime University, Gdynia, Poland

ABSTRACT: The paper presents some problems of carrying out measurements of energetic characteristics and vessel's performance in the conditions of sea examinations. We present the influence of external conditions in the change of vessel's hull resistance and propeller characteristics as well as the influence of weather conditions in the results of examinations and characteristics of gas turbine engine. We also discuss the manner of reducing the results of measurements to the standard conditions. We present the way of preparing propulsion characteristics and the analysis of examination uncertainty for the measurement of torque.

1 SYMBOLS

B –fuel consumption [kg/h];
b_e – unit fuel consumption [kg/kWh];
G_K –air consumption [kg/s];
D – propeller dimension [m];
J – advance of propeller;
K_Q – torque coefficient;
K_T – thrust coefficient;
L – work [kJ];
M – mass [kg];
N – power [kW];
n – rotational speed of a propeller [1/s];
p – pressure [Pa];
Q – torque [Nm];
R- Vessel resistance [N];
T – propeller thrust [N];
T – temperature [K];
t – temperature [°C];
t – suction coefficient ;
w –wake fraction;
x – content relative to dry air mass;
v_p – propeller advance speed [m/s];
ρ – water density [kg/m³];
η_p – freewheeling propeller efficiency;

Index's:
h – per hour;
m – concern measured parameters;
o – ambient parameters;
r – concern reduced parameters.

2 INTRODUCTION

Measurements performed on vessels are aimed at determining the up-to-date technical condition of the elements of the main propulsion or the evaluation of the operating elements of a vessel. The diagnostic measurements should be performed in a continuous manner, and the measurements to determine propulsion characteristics should be performed at specified time points, e.g. after completing the construction works on the vessel, after repairing elements of the propulsion system, etc. The measurements are performed on a vessel to develop propulsion forecast for a new built ship, or to evaluate current operating parameters of an exploited vessel. Irrespective of the aim of the measurements, it should be noted that a vessel always works in different conditions and the conditions may affect the quality and reliability of measurements. The change of the conditions for vessel movement is induced by parameters linked with:
– the vessel, i.e. vessel loading, use of reserves (change in displacement), change in the condition of hull, propellers, engines, etc.
– hydrometeorological conditions
– vessel operation region.
   The evaluation of factual propulsive characteristics in exploitation is performed during vessel sea trials [2].
   In order to fully evaluate the propulsive characteristics, the following should be measured: torque on propulsion shafts; propeller thrust; rotational speed of shafts, vessel speed and the use of fuel by particular engines. Fig. 1 presents a block diagram of a vessel as the object of sea propulsion trials.

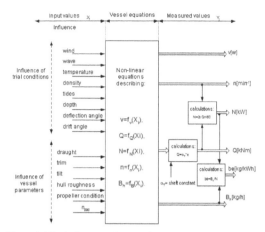

Figure 1. Block diagram of a vessel.

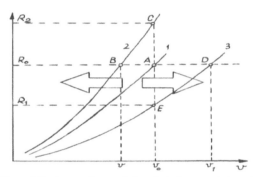

Figure 2. Resistance characteristics of a hull: 1. nominal ambient conditions (design); 2. degraded ambient conditions: 3. improved ambient conditions.

3 CHANGE IN VESSEL FLOATATIONAL RESISTANCE

Vessel resistance is determined at the design stage swith the use of computing methods and experimentally – with the use of model trials. They are the basis for selecting the propulsive system. At the design stage the resistance of a vessel is determined for standard navigational conditions. During exploitation, displacement, and consequently – draught, hull state, external conditions, etc. change continuously. This leads to a change (deterioration) in resistance characteristics and a change in the type of main engine load when the same vessel speed is developed. Thus, the information on resistance characteristics and the evaluation of the influence of particular conditions which, in turn, affect their values is significant in the diagnostic assessment of the state of propulsion system elements and of propulsive characteristics [3]. Fig. 2 illustrates exemplary resistance characteristics for a vessel operating in improved or worsened operating conditions.

4 CHANGE IN PROPELLER CHARACTERISTICS

Just like a hull, vessel propellers work in vastly varying conditions. It is especially applicable to changes in propeller draught resulting from displacement, permanent draught change and the angle of the incoming water during wave navigation, as well as deterioration in the condition of propeller blade surface (increased roughness). In order to evaluate the conditions in which a propeller operates at the rear of a vessel hull, it is important to know the cooperation relationship between the hull and propeller. Fig. 3 illustrates exemplary hydrodynamic characteristics of a propeller operating at the rear of a vessel hull.

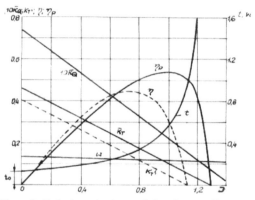

Figure 3. Hydrodynamic characteristics of a propeller: -------
free propeller characteristics in undisturbed water velocity field; - - - - characteristics of the working propeller after the ship's hull.

5 THE INFLUENCE OF EXTERNAL CONDITIONS ON ENGINE CHARACTERISTICS

The propulsive system of a vessel operates in vastly varying conditions. The change of conditions is caused by continuous change in displacement, and also draught, change of region where a vessel operates, change of hydrometeorological conditions, and changes in the condition of the hull, propeller and engines. In order to diagnose the propulsive system of a vessel in time, it is necessary to take changes in operating conditions into consideration.

5.1 The influence of atmospheric parameters

Atmospheric conditions affect the performance of each engine type, major influence, however, is observed in gas turbine engines [1]. In order to ensure adequate course of operation processes, gas turbine engines need considerable amounts of air. Excess air coefficient in the engine is 3,6–5. This accounts for unit air demand of 18 – 25 kg/kWh. The need for compressing large masses of air increases the importance of the influence of change in atmospheric conditions on engine functions, conditions for regulation, performance, etc. Significant influence is produced by changes in temperature, pressure and humidity of air, which cause changes in physical properties of the operating factor, such as density, viscosity, heat capacity, gas constant, etc.

Changes in engine performance resulting from atmospheric conditions may be considerable and sometimes may hinder the achievement of adequate engine performance, or the diagnosis due to the incomparability of measurement conditions.

5.1.1 *The influence of incoming air temperature*

Changes in incoming air temperature are due to the fact that vessels are exploited in various regions, or even climate zones, various seasons of the year, and day times.

The standard assumption is that ambient temperature is 288 K. And for the region of the Baltic Sea it may be assumed that ambient temperature fluctuates within the range of 238 –308 K. Such large fluctuations lead to considerable changes in engine work conditions, which needs to be taken into consideration while evaluating performance in an engine that operates in various conditions. The increase in incoming air temperature leads to reducing the air mass stream due to reduced density, and, as a result - decrease of engine power. What also changes are other figures that characterize the course of the working process of an engine and compressor efficiency. In the ranges of load that are close to those is calculations, the increase in air temperature leads to a minor increase in compressor efficiency. This is caused by an increase in sound speed and decrease of Mach number, as a result of which the conditions of transitional flow are improved, which translates to reduced hydraulic loss.

When incoming air temperature drops, the decrease of compressor efficiency leads to an increase in unit fuel consumption. Fig. 4 illustrates the properties of changes in compressor efficiency and its effective work depending on air temperature for various compression values. The presented relationships indicate that optimum compression is subject to linear changes both for compressor efficiency and work.

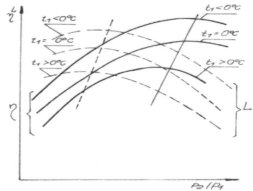

Figure. 4. The properties of changes in compressor efficiency and its effective work depending on air temperature and compression:
_____ The optimum range of efficiency;
------ The optimum range of effective work.

The larger the difference in temperature, the larger the differences in the changes of optimum values.

5.1.2 *The influence of atmospheric pressure changes*

In comparison with temperatures, changes in atmospheric pressure are relatively minor. Changes in air pressure may be within the range 96 –104 kPa. Relative change of pressure in relation to standard pressure (101,3 kPa) is up to 10%. That is why the influence of pressure change on the properties of engine functions is not as significant, as the influence of temperature. Change of air pressure and the resulting change in air density at the engine inlet leads to proportional changes in all engine control cross-sections. An increase in atmospheric pressure leads to increasing air mass and, as a result, increase in engine power. What does not change is temperature, rotational speed, compression, efficiency and unit fuel consumption.

5.1.3 *The influence of change in air humidity*

Air humidity may be subject to a wide range of changes – from dry air to air containing saturated vapour. Humidity indeed affects gas engine performance. It is especially related to changes in air mass and with changes of air heat parameters, such as heat capacity and gas constant. An increase in humidity leads to an increase in gas capacity, leading to a decrease in incoming air density. That, in turn, leads to decreasing the volume of air flow through an engine. The influence of the decreased volume of air flow is larger than the increase in heat capacity, which leads to engine power drop. Apart from vapour, the incoming air also contains water drops in the form of sea spray. Moistening degree is determined based on water and vapour content relative to dry air mass.

$$X = \frac{m_{H2O}}{m_{ps}} \qquad (1)$$

where m_{ps} - dry air mass.

Fig. 5 illustrates an example of change in engine performance when the change in moistening degree is within the range 0,01 – 0,07

5.2 Calculating the measured values to the so-called model atmosphere

For changeable conditions during vessel engine exploitation, it is necessary to relate the test results to the so-called model atmosphere (po = 101,325 kPa and To =288,15 K).

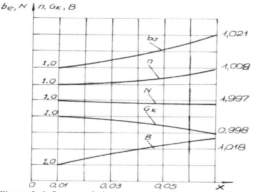

Figure 5. Influence of changes in incoming air humidity on turbine engine characteristics.

Changes in temperature, pressure, rotational speed and power relative to atmospheric conditions are presented in the following relationships:
– reduced engine power

$$P_{zr} = P_{zm} \frac{101325}{p_{ozm}} \sqrt{\frac{288,15}{T_{ozm}}} \qquad (2)$$

– reduced pressure

$$p_{zr} = p_{ozm} \frac{101325}{p_{ozm}} \qquad (3)$$

– reduced temperature

$$T_{zr} = T_{zm} \frac{288,15}{T_{ozm}} \qquad (4)$$

– reduced rotational speed

$$n_{zr} = n_{zm} \frac{288,15}{T_{ozm}} \qquad (5)$$

6 PROPULSIVE CHARACTERISTICS

In order to prepare propulsive characteristics, it is necessary to know the following:
– resistance characteristics of the hull R = f(v)..
– characteristics of freewheeling propellers
– characteristics of propulsive engines
– characteristics of elements transmitting the torque
Hydrodynamic characteristics of propellers in the form of K_T, K_Q, η_p = f(J)
where:
– thrust coefficient:

$$K_T = \frac{T}{\rho n^2 D^4} \qquad (6)$$

– torque coefficient:

$$K_Q = \frac{Q}{\rho n^2 D^5} \qquad (7)$$

– η_p – freewheeling propeller efficiency

$$\eta_p = \frac{K_T}{K_Q} \frac{J}{2\pi} \qquad (8)$$

– J - advance coefficient

$$J = \frac{\upsilon_p}{Dn} \qquad (9)$$

Coefficients which characterize hull and propeller cooperation
t – suction coefficient

$$t = 1 - \frac{R}{T} \qquad (10)$$

w –wake fraction

$$w = 1 - \frac{\upsilon_p}{\upsilon} \qquad (11)$$

The basis for propulsive characteristics is determining the area of possible operation for a freewheeling propeller on the grounds of hydrodynamic characteristics. The area is determined in coordinate systems T – n, Q – n, N – n with indicated lines of constant values of advance coefficient J and rotational speed of a propeller n. Next, resistance characteristics and propulsion engine characteristics are transferred onto adequate graphs and collated with the same measurement sites, e.g. propeller cone or output shaft clutch for torque and power, and vessel hull or propeller cone for resistance characteristics and propeller parameters. Collating the measurement results with appropriate sites is significant in order to consider the efficiency of particular elements that take part in transferring torque, and the efficiency of the hull and propellers. Propulsive characteristics offer the full presentation of the regularities in propulsion system element selection and make it possi-

ble to evaluate operating properties of a vessel. For vessels with combined propulsive systems, propulsion characteristics and the way in which they are presented are far more complicated. This is because:
- a combined propulsion system provides a number of ways to use diesel engines, e.g. propelling jet engines working on their own, peak engines working on their own, or both engine types working jointly;
- high navigation velocities result in high propeller strain; propellers usually work under highly developed cavitation, or supercavitation, hence, in addition to the advance ratio, the hydrodynamic characteristics of propellers must allow for the cavitation number.

Good results are achieved while presenting propulsion characteristics of combined systems as individual ones [4].

Fig. 6 illustrates an example of propulsive characteristics for classic vessel propulsion.

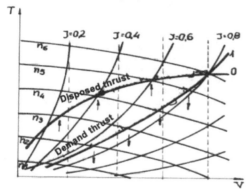

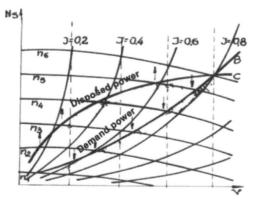

Figure 6. Propulsive characteristics of a propeller.

In order to properly evaluate propulsion characteristics achieved during sea trials, it is significant to estimate measurement uncertainty ranges for the measured and calculated values. Among the measured values, the largest measurement uncertainty is in measuring propeller thrust and torque.

Both values are measured by means of tensometry, with the use of contactless signal transmission from a rotating shaft.

The scope of measurement uncertainty for torque and thrust is mainly affected by the uncertainty of the evaluation of G shaft material resilience, which can be up to 4% and the error resulting from the failure to maintain parallel position relative to the axis of the shaft with tensometers, when propeller thrust is measured.

$$\frac{\partial Q}{Q \partial G} = \frac{1}{G} \frac{\partial Q}{Q \partial D} = \frac{3}{D} \frac{\partial Q}{Q \partial \varepsilon} = \frac{1}{\varepsilon} \qquad (12)$$

The uncertainty for the measurement of torque measured by means of tensometry consists of two fraction components.
- u_1 standard uncertainty of the measuring apparatus
- u_2 standard constant uncertainty of the shaft α_T.

$$u_Q{}^2 = \left(\frac{u_Q}{Q}\right)^2 = \left(\frac{u_G}{G}\right)^2 + \rho\left(\frac{u_D}{D}\right)^2 + \left(\frac{u_\varepsilon}{\varepsilon}\right)^2 \qquad (13)$$

The uncertainty of the measuring apparatus depends on the gauge used. Their borderline measurement error is 0,5%, thus the measurement uncertainty is

$$u_1 = \frac{0,5}{\sqrt{3}} = 0,289\% \qquad (14)$$

with the assumption of even distribution.

For tensometric models of a measurement system a calibrated resistor R_{cal} is used; its borderline error is 0,01%. The measured stress of shaft ε are in the following relation to the model:

$$\varepsilon = \frac{1}{4K_t} \frac{R_t}{R_t + R_{cal}} \qquad (15)$$

where:
R_t - tensometer resistance
K_t - tensometer constant

The influence of the model on measurement uncertainty for stress ε is determined according to relations like those for combined measurements

$$\frac{\partial \varepsilon}{\varepsilon \partial R_t} = \frac{1}{R_t}, \ \frac{\partial}{\varepsilon \partial K_t} = -\frac{1}{K}, \ \frac{\partial \varepsilon}{\varepsilon \partial R_2} = -\frac{1}{R_2} \qquad (16)$$

$$\left(\frac{U\varepsilon}{\varepsilon}\right)^2 = \left(\frac{V_{R_t}}{R_t}\right)^2 + \left(\frac{-U_k}{K_t}\right)^2 + \left(\frac{-V_{R_L}}{R_2}\right)^2 \qquad (17)$$

With borderline errors of basic values $R_t \rightarrow \pm 0,2\%$, $K_t \rightarrow \pm 0,5\%$ and $R_2 \rightarrow \pm 0,01\%$, the uncertainty for models of shaft stress is 0,314%. Torque is calculated based on the variable shaft stress value

$$Q = \alpha T \varepsilon \qquad (18)$$

where

$$J = \frac{\pi D^4}{32} \quad \alpha_T = \frac{4GJ}{D} \qquad (19)$$

α_T – constant for a praticular shaft
J- moment of inertia; G- shear modules

Torque measurement uncertainty calculated like it is the case in combined measurements and with borderline errors
for G 3s = 3,45 % thus $u_G / G = 1,15\%$
for D ±0,1 mm D>200 mm and $u_{D/D} = 0,029\%$
for ε $u_\varepsilon / \varepsilon = 0,314\%$ is: *1,2%*

Measurement uncertainty is largely due to errors in estimating G resilience module. This error may be eliminated by running resistance tests on steel used for shafts or using special scale measurement middlebody in the shaft line segment

7 CONCLUSIONS

Vessel operating system trials conducted in real conditions, with effects affected by external factors, i.e. trial environment, broadly understood hull and propeller condition may enable estimation of the technical condition of the whole vessel, i.e. hull and propulsion system. Periodical tests make it possible to determine reciprocal relationships between fuel use, torque, rotational speed, and vessel speed. These relations may be used in ongoing exploitation in order to evaluate the condition of particular elements of a propulsion system while using theoretical propulsion characteristics calculated for the adequate hull and propeller. In diagnostic tests the following factors need to be taken into consideration every time: vessel loading, for warships – reserves, including fuel reserves, which account for a considerable share of the total mass of the vessel, as well as atmospheric and hydrometeorological conditions of measurements.

REFERENCES

Charchalis A., 1991.Diagnozowanie okretowych silników turbinowych. wyd. AMW Gdynia.

Charchalis A., 2001. Nadzór eksploatacyjny siłowni z turbinowymi silnikami spalinowym.i PROBLEMY EKSPLOATACJI nr 4/2001

Charchalis A., 2001.Opory okrętów wojennych i pedniki okretowe. wyd. Akademii Marynarki Wojennej w Gdyni.

Charchalis A., 2003. Wykorzystanie charakterystyk napedowych układów ruchowych okretów szybkich. XXX Sympozjum DIAGNOSTYKA MASZYN. Wegierska Górka.

Charchalis A., 2004. Diagnozowanie układów napedowych okretów w oparciu o pomiar parametrów eksploatacyjnych. XXXI Sympozjum DIAGNOSTYKA MASZYN.Wegierska Górka

Charchalis A., 2006. Conditions of drive and diagnostic measurements during sea tests. Journal of KONES. Warszawa

13. Determination of Ship's Angle of Dynamic Heel Based on Model Tests

W. Mironiuk & A. Pawlędzio
Polish Naval Academy, Gdynia, Poland

ABSTRACT: Results of initial research on air flow's dynamic impact on a ship model of 888 project type are presented in the elaboration. The research has been executed at a test stand located in the Polish Naval Academy. The ship model of 888 project type has been an object of the tests. Results of executed measurements have been compared with theoretical calculations for an angle of dynamic heel. Input parameters for the tests and calculations have been defined in accordance with recommendations of Polish Register of Shipping (PRS) and IMO (IMO Instruments 1993, 2008, PRS 2007). Determination of a heeling moment by wind operation has been a key issue. The executed research has revealed that the way the criteria of the ship's dynamic stability are defined by PRS and IMO takes a certain safety margin into account.

1 INTRODUCTION

During operation, a ship is open to dynamic effects resulting from specifics of marine environment. Among these effects, wind dynamic impact together with a wavy motion is especially hazardous for the ship. Therefore, there are applicable criteria, taking the above said situations into consideration, in classification societies' regulations. A way to calculate a heeling moment originated by the wind operation is given in dynamic stability criteria. One may determine an angle of dynamic heel based on a function of this moment and righting levers' curve. But, assumptions and simplifications accepted by the classification societies result in omission of some phenomena and the angle of dynamic heel, determined based on theoretical calculations, differs from the real one. Determination of a real value of the dynamic angle of heel is possible by executing model tests performed with geometric and dynamic similarity scale taken into consideration. Results of theoretical calculations and measurements of dynamic angle of heel, executed with a ship model of 888 project type as an example, are given in this elaboration (Lewandowski, 2010). Results of the presented tests allow formulating practical conclusions which may be used by the classification societies, among the others.

2 RESEARCH FACILITY AND OBJECT OF THE TEST

Tests on the wind dynamic impact on a ship have been executed at a facility located in the Polish Naval Academy. Main elements of the stand are as the following:
- model basin for surface ships of internal dimensions of LxBxH 3x2x0,5 m,
- ship model of 888 project type,
- ship model of 660 project type,
- computer registering parameters of the ship model position,
- device generating air flow.

There is a set of fans fixed in a casing making a jet of a changeable section on the basin's shorter side. Dimensions of the jet's outlet are sufficient to let the air flow operate entire flank of the model, with a suitable reserve. There are 5 big and 5 small fans in the jet's housing – all together, they are capable of generating air flow faster than 5 m/s.

A ship model of 888 project type has been the object of the study. It was made in a scale 1:50 in respect to a real vessel. Its basic data have been as the following:
- overall length of the model: Lc = 1,444 m;
- length between perpendiculars: Lpp = 1,284 m;
- displacement of the model: D = 13,15 kg;
- average draught of the vessel: T = 0,078 m;
- depth of the model's centre of gravity: Z_G = 0,096 m;

Fig. 1. Facility for tests on dynamic impact of wind

At the level of floatation line, openings – horizontally arranged – are made in the model's stem and stern frame. Rods limiting the ship's drift caused by the air flow operation may be placed in the openings. Such an arrangement results in defining a fixed axis of rotation of the ship model. In fact, position of the ship model's axis of rotation is not permanent because, it depends – among the others – on dynamics of the wind effect. Location of the ship model's axis of rotation is crucial to determine the heeling moment. The IMO's stability code (IMO Instruments, 1993, 2008,) provides with instructions that the heeling moment's lever should be calculated as a distance from a centre of the topside projected area to the centre of the underwater part of the hull's projection on the symmetry plane, or – approximately – to the vessel's half draught depth. In this connection, results of measurements of and calculations for the dynamic angles of heel, executed in compliance with the IMO's recommendations, shall differ.

A sensor of the heeling and trim angles registering the angles with an accuracy of up to 0,01 of degree is installed on the model. Signals from the sensor are transmitted by means of a cable to a computer. Influence of the cables connected to the model has been omitted because of their insignificant weight and section areas.

3 PROBLEM OF GEOMETRIC AND DYNAMIC SIMILARITY SCALE

Solving the geometric and dynamic similarity scale problem has consisted in meeting given conditions. The ship model was made in the 1:50 scale - that is why all geometric quantities could have been easily calculated. Values of the righting levers curve of the ship model also have been subject to the geometric similarity principle. The heeling moment should effect in proportion to the vessel's righting levers, i.e. ratio of the ship's righting levers maximum value and value of the wind affected heeling moment's lever should be the same for both the model and the ship, as below:

$$\frac{GZ_{\max o}}{l_{wo}} = \frac{GZ_{\max m}}{l_{wm}} = const \qquad (1)$$

where the „o" subscript refers to the vessel, while the „m" subscript refers to the model. Figure 2 makes a graphical mapping of the (1) dependence.

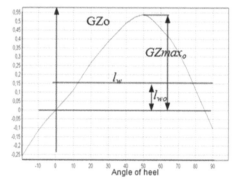

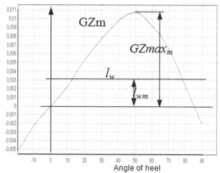

Fig. 2. Determination of wind affected heeling moment lever's value for ship model

Wind pressure affecting the vessel depends on the wind affected heeling moment. Therefore, the pressure should be defined in compliance with a suitable dynamic scale in relation to the model. Value of the pressure having dynamic impact on a real object, i.e.

on the vessel has been accepted in accordance with IMO and PRS regulations (IMO Instruments,1993, 2008, PRS 2007). It is 504 Pa for ships of unrestricted operational water areas and wind operating statically. However, a value 1,5 times bigger, i.e. 756 Pa is given for dynamic wind. A given wind velocity corresponds with this pressure value, and the velocity may be determined with many ways. Using dependence on the dynamic pressure is one of them, as the following:

$$p = \frac{\rho v^2}{2} \qquad (2)$$

The pressure value of 756 Pa gives the air speed of 35 m/s – at the density of 1,2 kg/m^3. To determine the wind velocity in respect to the pressure, one may also use table values provided in published references (Wiliński & Siemianowski 1993). It follows from them that the velocity should be some 29 m/s for squall of 756 Pa pressure.

The air speed obtained for the ship should be recalculated for a speed for the model. Suitable recalculations may be done with many ways. Using Euler coordinate is one of them (Grobyś 1998), as below:

$$Eu = \frac{p}{\rho v^2} \qquad (3)$$

The pressure of 756 Pa and the speed of 32 m/s both defined for the ship gives the wind velocity of 4,52 m/s, i.e. the velocity that should have impact on the model.

In case the following dependence on the wind affected heeling moment is applied (PRS 2008):

$$M_w = 2 \cdot 10^{-5} \cdot F_w \cdot z_w \cdot v_w^2 \cdot \cos^2\varphi \qquad (4)$$

where F_w – topside projected area [m^2], z_w – distance from centre of wind projected area to waterline positioned at height of T/2 above basic plane, at given load condition [m], φ – angle of heel, v_w – wind velocity at height of wind projected flank's geometric centre, defined with the following formula (PRS 2008):

$$v_w = v_{10}\left(\frac{z_w}{10}\right) [\text{knots}] \qquad (5)$$

where v_{10} – wind velocity at height of 10 metres above waterline; v_{10}=80 knots is accepted for the ships of unrestricted operational water areas. The wind velocity is 4,51 m/s for the ship model.

The presented solutions for the problem of dynamic similarity scale follow to similar results of the air flow speed. Therefore, it is probable that the calculations for the wind velocity have been done correctly.

4 EXECUTION OF THE TESTS

The study's programme has been executed in several stages. Determination of the air flow speed distribution has been one of the first activities at the research stand. Measurements of the air flow speed have been executed in 18 points. Results of the measurements for the speed of the air flow generated only by the big fans are given in Table 1. Average value of the speed of the air flow affecting the model has been 4,52 m/s.

Table 1. Results of measurements of air flow speed

Measurement height [cm]	Place of measurement and value of speed[m/s]						Average value [m/s]
	1	2	3	4	5	6	
35.5	4.57	4.68	4.69	4.16	4.10	4.53	4.46
18.5	4.67	4.86	4.77	4.53	4.16	4.65	4.61
8.5	4.33	4.63	4.60	4.65	4.65	4.10	4.49
							4.52

Next stage of the research has been to determine the dynamic angle of heel. Tests at the stand have been executed, among the others, for the following values of the angles of heel towards the windward shipboard: 6°, 15°, 18°. Values of these angles result from calculations of weather criteria executed for the ship project of type 888 in accordance with the regulations of IMO and PRS.

The fans have worked with constant velocity during registering the angles of heel what corresponds with constant characteristics of the heeling moment. Results of the measurements of the registered angles of heels are given on Figure 3.

a)

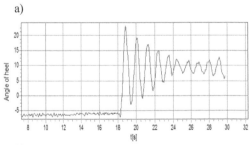

b)

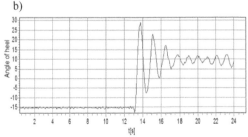

c)

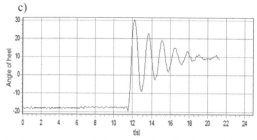

Fig.3. Measurements of dynamic angle of heel after deflecting model towards windward shipboard to angle of: a) 6°; b) 15°; c) 18°.

Table 2 contains a listing of the results for the measurements executed at the stand. The biggest values of the dynamic angle of heel have been obtained when the model has been deflected to the angle of 18 degrees towards the windward shipboard.

Table 2. Values of dynamic angles of heel

No.	1	2	3
Angle of heel towards windward shipboard	-6	-15	-18
Dynamic angle of heel	23	29	31

5 THEORETICAL CALCULATIONS

Calculations for the ship's dynamic angle of heel have been executed for the heeling moment defined in accordance with the IMO and PRS recommendations. Based on them, lever of the dynamically operating heeling moment has been determined assuming that the distance of the topside projected area's centre was measured from the half draught depth. The prospected lever has been calculated with the following dependence (IMO Instruments, 1993, 2008, PRS 2007)

$$l_w = 1.5 \frac{q_v F_w Z_v}{1000 \, g \, D} [m] \qquad (6)$$

where: q_v = 504 Pa – wind pressure; F_w- topside projected area [m^2]; Z_v- measured perpendicularly, distance from centre of topside projected area to centre of underwater part of hull's projection on symmetry plane, or – approximately – to vessel's half draught depth [m]; D – ship's displacement [t]; g – 9,81 m/s2; and for the data:

F_w =533 m2;

Z_v = 6.46 m;

D =1643.7 t,

to result in obtaining value of the wind affected heeling lever equal 0,162 m.

For this value, the dynamic angles of heel have been read on a graph (Fig. 4) and placed in a Table 3.

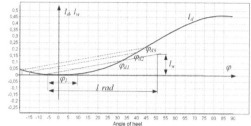

Fig.4. Determination of dynamic angles of heel for ship of 888 project type – for Z_v=6.46 m

Table 3. Values of dynamic angles of heel

No.	1	2	3
Angle of heel towards windward shipboard [deg]	-6	-15	-18
Dynamic angle of heel [deg]	33	40	43

The results show that the values of the dynamic angles of heel obtained from the calculations are seriously bigger than the values obtained from the measurements. The fact that lever on which force of the wind pressure operates was defined from the ship's half draught depth, instead of from the operative waterline, is one of the reasons. Should one take this note into consideration, the wind affected dynamic lever l_w shall equal 0.111 m. A next graph, given on figure 5, allowing defining the dynamic angles of heel, has been executed for this case. Obtained values of the angles are presented in a Table 4. Moreover, a line with values of the dynamic stability angles measured at the stand has been added - to enable to compare the results.

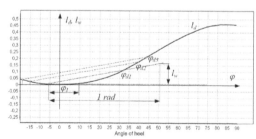

Fig.5. Determination of dynamic angles of heel for ship of 888 project type – for Z_v=4.49 m

Table 4. Values of dynamic angles of heel

Lp.	1	2	3
Angle of heel towards windward shipboard [deg]	-6	-15	-18
Dynamic angle of heel determined from figure 5 [deg]	25	32	36
Dynamic angle of heel measured at stand [deg]	23	29	31

This time, the results of the dynamic angle of heel calculations and measurements are very similar. Differences are from 10 to 16%. It is possible to obtain more similar results after damping of movement and size of the wind exposed area's projection, which changes during the heel, are taken into account.

6 SUMMARY

The executed initial research on the dynamic influence of the wind affected heeling moment shows high convergence of the theoretical calculations with the measured values. The obtained results prove that the way the wind affected heeling lever is determined has significant impact on the values of the dynamic angles of heel. The ship model has had the fixed axis of rotation, positioned in not waving water plane. Therefore, it has not been possible to compare the measurements results with the results of calculations for the dynamic angle of heel accurately, if determined based on the heeling moment imposed by IMO. Further research on the described issue shall allow executing more accurate analysis.

REFERENCES

IMO Instruments 1993. Code on Intact Stability of All Types of Ships. London.
IMO Instruments 2008. International Code on Intact Stability: London.
Gryboś R. 1998. Podstawy mechaniki płynów część 2. PWN.: Warszawa.
Lewandowski G. 2010. Badania modelowe stateczności dynamicznej okrętu projketu 888 poddanego oddziaływaniu wiatru- praca dyplomowa. AMW.: Gdynia
PRS 2007. Przepisy klasyfikacji i budowy statków morskich, Część IV. Gdańsk
PRS 2008. Przepisy klasyfikacji i budowy okrętów wojennych, Część IV. Gdańsk
Mironiuk, W. 2006. Preliminary research on stability of warship models. COPPE Brazil: Rio de Janeiro.
Pawłowski, M. 2004. Subdivision and damage stability of ships Gdańsk.
Wiliński A., Siemianowski T. 1983.: Teoria, konstrukcja i OPA okrętu. Zbiór zadań z teorii okrętu. WSMW.: Gdynia

14. Propulsive and Stopping Performance Analysis of Cellular Container Carriers

J. Artyszuk
Maritime University, Szczecin, Poland

ABSTRACT: The present contribution reviews all fundamental correlations between ship main particulars for fully cellular container carriers as based on a deep statistical analysis of world fleet. The use is made of exported data from the Lloyds Register of Ships ship internet database 'Sea-web' and the author's dedicated data-processing algorithms. Those parameters include ship size related items (length, displacement, deadweight, gross tonnage, trade-specific capacity dimensions ie. TEU), ratios of main dimensions, speed and main propulsion power. The research is primarily aimed at providing some necessary data to ship designers, waterway engineers, and ship operators engaged in port or fleet development projects, where navigation safety and efficiency is of utmost importance (like eg. a determination of the so-called characteristic and/or extreme ship within respective limits of ship size). The presented relationships certainly affect ship resistance-propulsive performance, seakeeping and manoeuvring characteristics. As an example, a simple but rather powerful mathematical model of generic type for ship stopping behaviour in coasting and crash regimes (taking place in ocean or harbour manoeuvring) is worked out.

1 INTRODUCTION AND GEOMETRY

The scientific interest on ship stopping performance has got a long history and the latest focus - refer eg. to Clarke & Hearn (1994), Hooft & van Manen (1968), Kynast (1981), Okamato et al. (1974), Sung & Rhee (2005). The following study significantly contributes to this area at least through the fully statistics based approach.

According to the Sea-web database (LR/IHS 2011) the population of fully cellular containers carriers (marked with the symbol 'A33A2CC', as based on the IHS Fairplay - Statcode 5, ver. 1075) consists of nearly 5500 ships. Among the available items/fields of interest for propulsive/manoeuvring analysis, only the ship's displacement m is not fully covered (30% of values are missing), that is of major importance itself and contributes for example to the calculation of block coefficient. However, for more detailed studies this lack of data can be estimated/overcome (with due regard of potential inaccuracy) taking into account various correlations with other size-dependent parameters (eg. deadweight, *TEU*, or gross tonnage).

Figures 1-5 present useful cross-dependence (and scattering) of different geometric parameters in case of cellular container carriers. The base for these plots (and all other figures in the paper) is the ship's length L meaning the length between perpendiculars

L_{BP}. This parameter is almost fundamental for ship design and performance analysis, though in harbour engineering and navigation safety studies the ship's length overall is often applied (LOA). Within the fleet of cellular container ships the ratio LOA/LBP is in the range (1.04, 1.06) for large size vessels and (1.05, 1.10) for smaller crafts. The average value is 1.06. The block coefficient cB computed from the ship's displacement and main dimensions (LBP, moulded breadth B, summer draught T) is between 0.58 and 0.7 (0.66 on the average). The ship's displacement to deadweight ratio starts from ca. 1.3 and reaches 1.4 (even 1.5 for smaller vessels), 1.36 is the mean value.

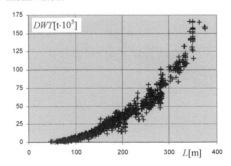

Figure 1. Deadweight vs. ship's length (c/c).

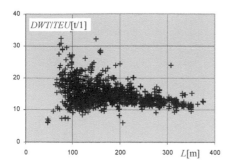

Figure 2. DWT/TEU ratio vs. ship's length (c/c).

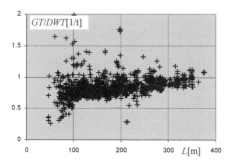

Figure 3. Gross tonnage/DWT ratio vs. ship's length (c/c).

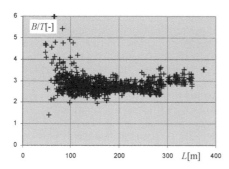

Figure 4. Ship's breadth/draught ratio vs. ship's length (c/c).

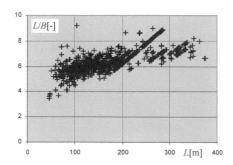

Figure 5. Ship's length/breadth ratio vs. ship's length (c/c).

It is apparent from Figure 2, illustrating the deadweight tons divided by TEU capacity, that there is a considerable spread of TEU arrangement among ships up to 200m in length. Figure 3 confirms somehow constant quotient of gross tonnage value to deadweight (approx. 0.83). Figures 4 and 5 show the average beam-draught ratio 2.7 and length-beam ratio 6.6, though within the latter values one must realize the higher variation ca. ±1.2 (as compared to ±0.3 in B/T). The easily noticeable in Figure 5 linear trends of L/B (for ships over 200m) probably constitute the design trends of particular shipyards.

2 PROPULSION ANALYSIS

The cellular container carriers are equipped by rule with a single screw propeller and a single main engine (both 99% of the whole population). The screw propeller can be either of fixed pitch (FPP, 83%) or controllable pitch type (CPP, 17%). The propulsion transmission is achieved via a direct drive (83%) or a geared drive (mechanical or electric, the remaining 17%). However, the FPP and direct drive (in view of accidentally the same above percentages and easily arising wrong illusion) may not methodologically and technically be linked to each other. Having a further insight to the statistics, in case of FPPs the direct drive (equal engine and propeller rpm) is strictly applied on 96% ships, while for CPPs the dominating is geared drive (78%).

Figures 6-9 provide additional propulsion related information. The smaller ship the lower is the service speed v_{serv}(Fig. 6), though a definite limit of 25kt is occurring for the largest ships (probably being quite sufficient from the economic point of view). The 'potential or allowable' speed increase with higher ship's length is well explained in the next Figure 7 presenting the nondimensional Froude number, $F_{nL}=v_{serv}/\sqrt{(g \cdot L)}$, which is hydrodynamically responsible for a rapid hull resistance increase at the higher speed due to the wave-making resistance component, and thus the power demand or fuel consumption. The Froude number is kept almost constant within the range from 0.2 to 0.3, but for larger ships the scattering is much narrower around 0.25.

Since the effective horse power (*EHP*), as affecting the brake horse power (*BHP*) in some way, is theoretically/roughly proportional to the square of ship's length and the third power of speed, the best evaluation method of the ship's BHP power installed (and reported) is to introduce the following 'relative power index':

$$HP^* = \frac{0.5\rho L^2 v_{serv}^3}{BHP} \qquad (1)$$

where ρ = water density (1.025t/m³); L[m]; v_{serv}[m/s]; BHP[kW].

The relative power index can be connected with the inverse of hull resistance coefficient, but already after incorporating the total propulsion efficiency. Precisely, It defines in rather general terms the ratio of power demand to the actually installed power. The higher value the less horsepower margin should be expected or the ship's hull fairing is more propulsively (hydrodynamically) effective. The similar relationship to Equation 1 exists in the well known Admiralty formula. It is evident from Figure 8, and better proved in the next paragraphs, that hulls of larger ships are more effective (the second option - an extra power is rarely the case).

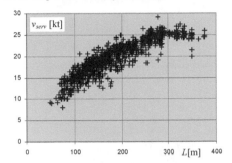

Figure 6. Ship's service speed (c/c).

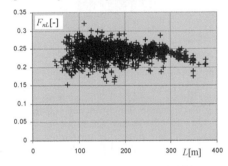

Figure 7. Froude number vs. ship's length (c/c).

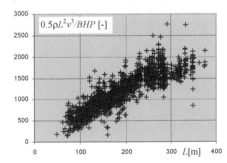

Figure 8. Index of relative horse power vs. ship's length (c/c).

Figure 9 gathers revolutions per minute ('rpm', though in the subsequent paragraphs the term 'rpm' is used in a wider sense as denoting the general propeller/engine rotative speed, the unit of which de-

pends upon the particular application/purpose) of the main engine, as selected for all direct drive ships. The collection comprises 79% of ships (among others due to some missing data in rpm). Only for a direct drive, the same rpm will be experienced both by the engine and propeller, that allows to use them in our propulsion/stopping analysis. It is namely common that the propeller rpms (as fundamental in the aforementioned analysis) are rarely reported in and able to be extracted from electronic databases. The latter are generally designed/purposed for economists and managers rather than for more technical studies.

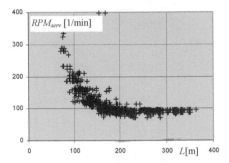

Figure 9. Main engine (direct drive propeller) service rpm vs. ship's length (c/c).

The rpm given in Figure 9 is the nominal main engine rpm but converted to the usual main engine service output (the so-called operational or engine margin is to be applied). In the study the level of NCO/MCO is assumed at 80%. The recalculation is according to the basic design principles of the main engine working point in normal service conditions - 100%MCO, equal to BHP, shall be achieved at nominal rpm, and of course at higher ship's speed than the above reported service speed (here to be linked with just the service rpm):

$$n_{serv} = \sqrt[3]{0.8n_n} = 0.928n_n \qquad (2)$$

where n_{serv} = service rpm (engine/propeller rotative speed); n_n = nominal rpm.

Since any propulsive/stopping behavior is governed by the hull and propeller forces, the unknown geometric and hydrodynamic parameters of the propeller and hull have to be identified. This task could be accomplished using the well known propeller early design diagrams. For the purpose of present study the work of van Lammeren et al. (1969) has been taken the advantage. One should refer to the original curves for optimum diameter of B4.70 propeller series (fig. 29, pp. 288 in the cited reference) as provided for the full-scale propeller Reynolds number of order $5 \cdot 10^7$ in case of container ships, though in future one could attempt to use charts for 5-bladed propellers (as often encountered on container vessels).

This backward (or hindcast) identification, additionally assuming a default constant wake fraction $w = 0.3$ and thrust deduction factors $t_p = 0.15$ (relative rotative efficiency is kept equal to unity), has revealed the results displayed in Figures 10-14 (propeller diameter D, propeller pitch ratio P/D, propeller open-water efficiency η_P, propeller advance ratio J).

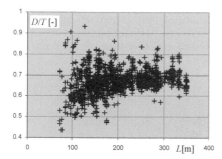

Figure 10. Propeller diameter/ship's length ratio vs. ship's length (c/c).

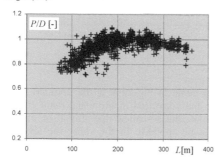

Figure 11. Propeller pitch ratio vs. ship's length (c/c).

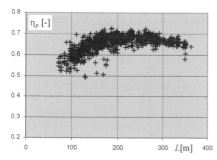

Figure 12. Open-water propeller efficiency vs. ship's length (c/c).

Figure 10 just shows a guess of propeller diameter in the non-dimensional way as a fraction of ship's draught. The ship's hull resistance coefficient, as referenced to the ship's length and draught product LT has been computed by the following formula:

$$c_R = \frac{\left(1 - t_P\right)\rho n_{serv}^2 D^4 k_T(J)}{0.5\rho LT v_{serv}^2} \tag{3}$$

where k_T = propeller thrust coefficient.

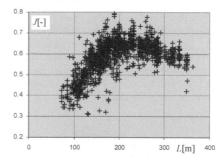

Figure 13. Propeller advance ratio vs. ship's length (c/c).

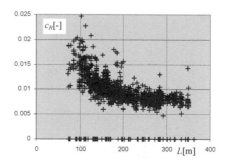

Figure 14. Ship resistance coefficient vs. ship's length (c/c).

The latter thrust coefficient k_T has been established based on the propeller efficiency η_P, the advance ratio J (Figs 12,13), and the propeller torque coefficient k_Q, which in turn, after the former identification of J, can be calculated from the Taylor parameter B_P, being the input to the aforementioned propeller design charts (the B_P parameter is the limiting combination of J and just k_Q that ensure the best efficiency within the available horsepower, rpm, and ship's speed).

It is clear (Fig. 14) that hull designs for larger ships are considerably highly effective, and that from a certain ship size this effectiveness (in term of hull resistance coefficient) is almost constant.

3 COASTING STOP GENERAL MODEL

The ship coasting stop behavior can be well modeled in a single dimension (1DOF), just describing the forward straight-line movement using only the hull force in forward direction ie. the ship's resistance. All other forces can be neglected, among which there are very low in magnitude propeller forces arising in turbine modes or even shaft brake modes.

The yaw motion and its contribution (together with the hull drift angle) to the also braking inertia forces of centrifugal and Coriolis nature is not huge in this kind of maneuver. The basic equation is now as follows:

$$m(1 + k_{11})\frac{dv}{dt} = 0.5\rho LTv^2 c_R(v) \tag{4}$$

where k_{11} = added mass ratio (assumed 0.05).

The dependence of c_R on v is essential for high Froude numbers and low Reynolds number. For the usual range of Froude number (covering ship's speeds from low to high), the resistance coefficient c_R can be considered constant (resistance-speed square relationship), but for very low ship's speeds c_R shall accept exponential relationship leading to the well known linear dependence of resistance (drag) on the ship's speed. The very low speed range is not the case in the present study, thus c_R is constant, and can be assessed eg. as per Figure 14.

Since $m = \rho c_B LBT$, one can arrive at:

$$\frac{dv}{dt} = \frac{0.5\frac{L}{B}c_R}{c_B(1+k_{11})}\cdot\frac{v^2}{L} = A\cdot\frac{v^2}{L} \tag{5}$$

where A = coasting stop parameter (non-dimensional, negative due to negative c_R).

The dimensional Equation 5 can be solved analytically. It is not comfortable for an analysis of a huge selection of ships, different in size and speed - the stopping/deceleration time t for a given input speed drop Δv will be a function of the approach speed v_0 and ship's length:

$$\left[\frac{1}{v_0 - \Delta v} - \frac{1}{v_0}\right] = -\frac{A}{L}t \tag{6}$$

where v_0 = coasting stop parameter (non-dimensional, negative due to negative c_R).

It is absolutely better (i.a. from the navigation and/or shiphandling practice point of view, where the distance rather than time is a safety/reference factor) to directly introduce the traveled distance s.

Finally, after adopting the non-dimensional distance $s' = s/L$ (ie. the distance expressed in ship's length units, which is equivalent to the so-called non-dimensional time $t' = t'\cdot L/v$), one can read:

$$\frac{dv}{ds'} = Ads' \tag{7}$$

$$\ln\left(\frac{v_0}{v_0 - \Delta v}\right) = \ln\left(\frac{1}{1 - \frac{\Delta v}{v_0}}\right) = -As' \tag{8}$$

Equation 8 states that the non-dimensional deceleration distance (the stopping distance until the re-

quested speed reduction is achieved) is independent of the ship's length and is only affected by the ratio of speed reduction to the approach speed. The latter means that performing coasting stop from 4 to 2 knots requires the same distance as with 10 and 5 knots respectively. However, in the latter case the stopping absolute time will be much shorter. One shall also realize here that a ship can not be brought down up to the 'dead in water' status with this simple model

For cellular container carriers in concern, the values of A parameter are gathered in Figure 15. The generic diagram in Figure 16, as based on Equation 8, is aimed to provide a prediction of non-dimensional distance - eg. for a speed reduction ratio $\Delta v/v_0$ of 0.4 (40% drop of the initial speed) and the A value of -0.05, one can reach 10 ship's lengths of stopping distance.

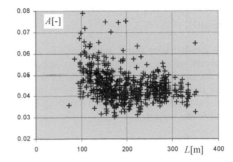

Figure 15. Coasting stop model parameter vs. ship's length (c/c).

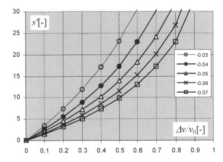

Figure 16. Coasting stopping non-dimensional distance vs. relative speed reduction - general model.

4 CRASH STOP APPROXIMATE MODEL

The ship dynamics in the crash stopping regime (disregarding the propeller lateral force, or the so-called paddle wheel effect, as exciting the drift/yaw combined motion, through which the stopping is more effective, to a certain degree) is governed by the hull force (refer to the previous chapter) and the propeller negative thrust:

$$m(1+k_{11})\frac{dv}{dt} = 0.5\rho LTv^2 c_R + (1-t_P)\rho n^2 D^4 k_T(J) \quad (9)$$

The thrust coefficient k_T in the crash regime quadrant ($v>0$, $n<0$, $J<0$) should be selected from the available propeller hydrodynamic characteristics for a given P/D ratio - only the FPP and the reversing diesel engine is discussed hereafter as the most critical propulsion system arrangement.

Equation 9 ultimately leads to the following:

$$\frac{dv}{ds'} = A \cdot v + \frac{(1-t_P)(1-w)^2 D'^2}{c_B(1+k_{11})\dfrac{B}{T}} \cdot \frac{k_T(J)}{J^2} \cdot v \quad (10)$$

$$\frac{dv}{ds'} = A \cdot v + B_1 \cdot \frac{k_T(J)}{J^2} \cdot v = Av + B_1 \frac{k_T^*(J^*)}{J^{*2}\dfrac{P}{D}} \cdot v \quad (11)$$

where $D' = D/T$; $k_T^* = k_T/(P/D)$; $J^* = J/(P/D)$; $J = v(1-w)/(nD)$.

Making a rather very rough assumption (a quantitative analysis of such an approach is scheduled soon) that the crash stop is performed under the propeller force only, one can arrive at the very convenient analytical model of ship's crash stopping (up to a ship dead in the water):

$$\frac{dv}{v} \cdot \frac{J^{*2}}{k_T^*(J^*)} = \frac{B_1}{\left(\dfrac{P}{D}\right)} ds' \quad (12)$$

Since $dv/v = dJ^*/J^*$ then

$$\int_{J_0^*}^{0} \frac{J^*}{k_T^*(J^*)} dJ^* = \frac{B_1}{\left(\dfrac{P}{D}\right)} s' \quad (13)$$

The left side integral of Equation 13 can be calculated off-line (it assumes a ship stopping from a certain initial advance ratio J_0^*, affected by the approach speed and the engine ordered negative rpm, up to the complete ship's stop) and constitutes an additional universal chart of propeller action, especially as independent from the pitch ratio. This models enables to catch at glance some characteristic relationships in ship's stopping behavior, according to which an optimization of ship design and operation can be carried out The concept of universal (independent from the pitch ratio) propeller force description and thus the universal chart of the thrust coefficient (referred to as the pitch or slip related coefficient, the respective symbols, also for the advance ratio, are marked with the asterisk) has been widely analyzed in Artyszuk (2001).

The full model for combined hull and propeller forces looks like:

$$\int_{J_0^*}^{0} \frac{J^* dJ^*}{AJ^{*2} + B_1 \dfrac{k_T^*(J^*)}{\dfrac{P}{D}}} = s' \quad (14)$$

in which the distance s' is unwillingly dependent on the pitch ratio.

Figure 17 encompasses B_1 parameter values for the class of cellular container ships. In Figures 18,19 (just for the reference purpose) are enclosed the aforementioned propeller characteristics quadrant for the crash regime, together with its transformation and the integral computation (Eq. 13).

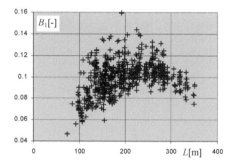

Figure 17. Crash stop model parameter vs. ship's length (c/c).

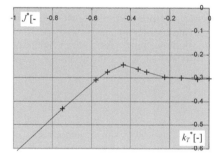

Figure 18. k_T-J diagram for the crash regime.

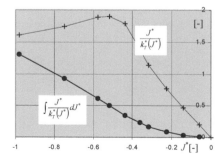

Figure 19. Left side integral parameter of general crash model.

While analyzing the crash stop with the FPP, it is widely known that the ordered negative engine rpm can not be executed (the engine is reversed by means of the start air) at rather high engine positive rpm that can still occur after the fuel cut off. The relatively high rpm are caused by the so-called turbine mode of the propeller operation - propeller inflow to the propeller from the fast moving ship (but slower and slower with time) maintains the rpm ensuring zero angle of attack of the propeller blade. The propeller torque, and thus the torque coefficient shall mathematically be equal to zero $k_Q = 0$ (no power demand), but the thrust coefficient k_T, and hence the thrust itself, is negative, though relatively small. In the turbine mode, the advance ratio J^*_{turb} (as independent from the pitch ratio) is equal to 1.1 (Artyszuk 2001) - the diagram $k^*_T(J^*)$ and the standard $k_T(J)$ diagram for $P/D=1$ are identical by definition - and leads to the following linear relationship between speed and rpm

$$n_{turb} = \frac{(1-w)}{J_{turb}D}v = \frac{(1-w)}{1.1\frac{P}{D}\cdot D'T}\cdot v \qquad (15)$$

The engine rpm, at which the start air is effective for reversing purposes, is called the reversing rpm n_{rev} and established by the engine's manufacturer. Often, the fraction of the studied before service rpm (n_{serv}) equal to 0.35÷0.40 is specified in this context - the latter value $n'_{rev}=\underline{0.4}$ is further used in the present simulation. This means one must wait for a ship's forward speed to go below the so-called ship's reversing speed v_{rev} (sometimes corresponding even to the speed developed under the SAH or HAH telegraph setting):

$$v'_{rev} = \frac{v_{rev}}{v_{serv}} = \frac{J^*_{turb}}{J^*_{serv}}n'_{rev} = \frac{1.1\cdot}{\left(\frac{P}{D}\right)}0.4 = 0.44\frac{\left(\frac{P}{D}\right)}{J} \qquad (16)$$

where one should refer to Figures 11, 14. The calculation results according to Equation 16 are depicted in subsequent Figures 20, 21.

The J_0^*, as based in the present study on the FAS rpm, can be adopted at the level of 80% n_{serv} (corresponding to the maneuvering FAH throttle, which also indicates the engine output power approx. 51% of NCO). Finally one can read:

$$J_0^* = J^*_{turb}\frac{n'_{rev}}{n_{FAS}} = J^*_{turb}\frac{0.4}{0.8} = J^*_{turb}\cdot 0.5 \qquad (17)$$

It can be concluded that for larger ships one should expect the higher reversing speed (even in the order of 15kt and more, compare with Fig. 6), while at the same time its ratio to the service speed

is nearly steady (0.7 on the average, just for reference $n'_{rev}=0.35$ contributes to v'_{rev} of order 0.6).

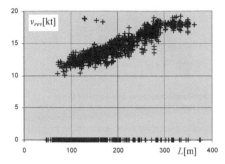

Figure 20. Speed for reversing the main engine (c/c).

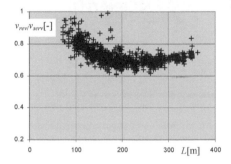

Figure 21. Reversing speed to service speed ratio (c/c).

However, there is a significant variation of the turbine rpm balance point among some full scale ships. It means that some of them reduce the engine rpm more rapidly than the others, for which the possible increase of the wake fraction shall be blamed - see Equation 15. Such an effect shall be paid a special attention in the future research.

In the final Figures 22-25 one can find the stopping distances in both regimes of crashing maneuver for container ships: the coasting stage part from the approach speed up to the reversing speed - s'_{coast}, using the model of Chapter 3, and the specific crash stage by means of this chapter approximate model - s'_{crash} (with the hull drag disregarded what really underestimates the stopping efficiency, in reality this distance shall be lower). All values are given in ship's length units, that can be easily compared with eg. the IMO standards (relatively high approach speed, the influence of turbine mode) or the practical rules of thumb known to experienced ship handlers (harbour maneuvering at medium and low speed). Both coasting and crash relative distances are added together in Figure 24.

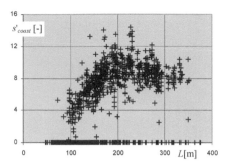

Figure 22. Coasting stopping distance (in ship's lengths) up to the reversing speed vs. ship's length (c/c).

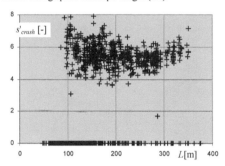

Figure 23. Approximate crash stage stopping distance (in ship's lengths) from the reversing speed vs. ship's length (c/c).

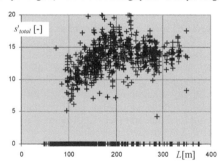

Figure 24. Approximate total stopping distance (in ship's lengths) till a ship dead in water vs. ship's length (c/c).

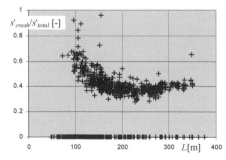

Figure 25. Crash stage part in the total stopping distance vs. ship's length (c/c).

The violation of IMO's standard 15 ship's lengths suggests a necessity of further improvement to the presented models and verification in full scale (from the standpoint of crash stop distance decomposition among the coasting and the proper crash phase, as expressed in ship's length units). One can namely believe that the violation is impossible in reality since the IMO standard is really a well established design (or verification) standard by the shipbuilding industry and known in some circles for its relatively mild requirements imposed upon the design process. In Figure 25 one can notice that the proper crash stage consumes nearly 40% of the total distance.

5 FINAL REMARKS

The performed research, beside some obvious shortcomings, has proved a great potential of the proposed methodology in analyzing ship stopping performance. Further developments and discussion in this particular area is recommended, where a proper feedback from all parties involved in the design and operation of ships is highly appreciated. The created generic analytical model of stopping behavior, after certain improvements, is also intended to be a benchmark for verification and tuning of other more sophisticated full mission models. It definitely sets an new standard in training navigators on ship handling as well, where a generic knowledge on ship maneuvering performance is often preferred over a specific behavior of a particular ship.

REFERENCES

Artyszuk, J. 2001. Propeller Slip Ratio in the Ship Manoeuvring Motion Mathematical Model - Thrust Case. In *Proc. of IV Navigation Symposium*, Maritime University, Gdynia, Jun 19-20 (CD-ROM).

Clarke, D. & Hearn, G.E. 1994. The IMO Criterion for Ship Stopping Ability. In *3rd International Conference - Manoeuvring and Control of Marine Craft (MCMC'94)*, Sep 7-9, Southampton: 9-19.

Hooft, J.P. & van Manen, J.D. 1968. The Effect of Propeller Type on the Stopping Abilities of Large Ships. *RINA Trans.* 110: 29-42.

Kynast, G. 1981. Some Approximate Formulas for the Acceleration and Stopping Manoeuvres. *Schiff&Hafen 33*(7): 65-72 (in German).

van Lammeren, W.P.A., van Manen, J.D. & Oosterveld, M.W.C. 1969. The Wageningen B-Screw Series. *SNAME Transactions 77*: 269-317.

LR/IHS 2011. Internet: www.sea-web.com (visited on Jan 30th, 2011).

Okamoto, H., Tanaka, A., Nozawa, K. & Tanaka A. 1974. Stopping Abilities of Ships Equipped with Controllable Pitch Propeller, Part I & Part II. *International Shipbuilding Progress* 21 (234&235 -Feb&Mar): 40-50 & 53-69.

Sung, Y.J. & Rhee, K.P. 2005. New Prediction Method on the Stopping Ability of Diesel Ships with Fixed Pitch Propeller. *International Shipbuilding Progress* 52(2): 113-128.

15. Coalescence Filtration with an Unwoven Fabric Barrier in Oil Bilge Water Separation on Board Ships

J. Gutteter-Grudziński
Maritime University of Szczecin, Poland

ABSTRACT: This article analyzes phenomena occurring in the process of oil-water emulsion flow through a coalescence fabric filter. The authors determined the flow of oil-water mixture through a fabric partition, the efficiency of filtration η_F and the distribution of oil particle size d_o by using a Malvern analyzer gauge. The results, presented graphically and in tables, can be considered as satisfactory as they meet the standards of Resolution MEPC 60/(33) IMO for shipboard equipment.

1 OIL SEPARATION ON BOARD SHIPS

Oily waters produced in ship operations are one of the essential threats to the marine environment. These wastes find their way to seawater by:
- discharge of ballast water and cargo tank washing water,
- discharge of bilge water,
- ship disasters and failures,
- spillage during oil un/loading operations,
- oil spills from propeller shaft bearings and other oil-lubricated elements that have contact with water [3,4,7,9,17].

All these sources of pollution have a high content of oil products, and their occurrence seems to be practically unavoidable. The oil separator NEPTUN shown in Figure 1 purifies bilge water in which oil particles reaching the oil separator filter can have diameters less than 50 μm and various oil concentration amounting to 10 000 ppm [4,24,25].

The diameter of smallest separated oil particle in the oil separation process on a fabric barrier is a function of many parameters [4]:

$$d_o = f\left(\rho_w, \rho_o, \eta_w, w, l, F, \varepsilon, g\right) \tag{1}$$

where: w – speed of mix flow [m/s], F – barrier surface area [m²], ε – coefficient of barrier porosity, ρ_w – density of continuous phase (water) [kg/m³], ρ_o – density of dispersed phase (oil) [kg/m³], g – Earth's gravity [m/s²], η_w – coefficient of water dynamic viscosity [N·s/m²], d_o – diameter of oil particle [m], l – track of oil particle (length of capilaries or pores) [m].

An analysis of oil/water emulsion found in a separate porous fabric barrier [4] showed that a diameter of the smallest droplets of oil separated in the filter can be determined as follows:

$$d_o \geq \left[\frac{4,5 \cdot q_v \cdot \eta_w \cdot d_w}{g \cdot V_c \left(1-\varepsilon\right)\cdot \left(\rho_w - \rho_o\right)}\right]^{0,5} \tag{2}$$

where: q_v – volumetric stream of water [m³/s], V_c – volume of sample filtration pad bed [m³], d_w – fiber diameter [m].

A conclusion we can draw from the above is that there are a few measures that can be taken to improve the efficiency of the filtration barrier capturing oil droplets due to coalescence:
- reduce q_v, d_w, ε and ρ_o;
- increase the inner surface area of the barrier;
- incrase fabric porosity;
- adjusting these flow parameters on an individual basis:
  - barrier parameters,
  - content of oil in the purified liquid.

2 ASSUMPTIONS FOR THE RESEARCH AND MODEL OIL SEPARATOR DESIGN

The research aimed to examine the efficiency of oil coalescence process on a new multi-layer barrier made of chemical fibres. A newly designed and constructed glass oil separator model was used for tests (Fig. 1). Glass fibres used so far in W-5 filters are now prohibited as they posed hazards to human health.

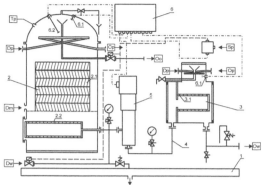

Fig. 1. Schematic diagram of the oil separator NEPTUN (q_v = 2.5 m³/h); 1 – base, 2 – gravity separator, 2.1 – corrugated plates unit, 2.2 – filter with a velours fabric, 3 – coalescence separator, 3.1 coalescence barrier, 4 – pipeline, 5 – pump, 6 – electric control panel, 6.1 – minimum oil level detector, 6.2 + maximum oil level detector; Dw – inlet of tap water, Dm – inlet of oil-water mixture, Tp – control thermostat, Dp – steam inlet, Op – steam outlet, Sp – compressed air, Oo – oil outlet, Ow – outlet of water of purified water

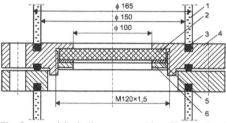

Fig. 2. A model of oil separator with a filtration barrier; 1 – glass cylinder, 2 – tested barrier, 3 – gasket, 4 – flange, 5 – washer, 6 – nut

Several assumptions were made before selecting the design solutions:
– the barrier will be horizontal, and oily water will be flowing upwards;
– oil will be stopped on the whole surface area of and inside the barrier, so that it separates in the form of droplets that will be gathering in the upper part of the cylinder;
– outflow of purified water from above the barrier through a connector pipe, mounted so that the collected oil will not get into it.

The oil separator model cosists of two glass cylinders (upper and lower) each closed by covers adequately squeezed onto them. To facilitate the exchange of unwoven fabric – the filtration barrier – there are easy-to-dimount flanges in the central part.

The intensity of emulsion flow through the filter has to be adjusted so that it corresponds to that in marine filter with D = 140 mm, H = 356 mm, working at q_v = 1 m³/h, assuming that the kinematic flow will be steady. As the same unwoven fabric barriers are used (the same capillaries) in the model as in marine filters, their Reynolds numbers will be equal:

$$\text{Re}_R = \text{Re}_M \qquad (3)$$

$$\left[\frac{w \cdot d_k \cdot \rho}{\eta}\right]_R = \left[\frac{w \cdot d_k \cdot \rho}{\eta}\right]_M \qquad (4)$$

where: (denotations see formula 1), indexes R and M refer to the marine filter and the lab filter.

From equation (4) we obtain:

$$\frac{\rho_R}{\rho_M} \cdot \frac{\eta_M}{\eta_R} \cdot \frac{d_{kR}}{d_{kM}} \cdot \frac{w_R}{w_M} = 1 \qquad (5)$$

Assuming that parameters ρ, η, d_k do not change, we get:

$$w_R = w_M = 1{,}77 \cdot 10^{-3} \text{ m/s} \qquad (6)$$

The volumetric stream in the model filter (Fig. 1, D = 100 mm) depends on the quotient of the barrier surface areas:

$$q_M = q_R \left(\frac{F_M}{F_R}\right) \qquad (7)$$

Based on the above considerations, we assumed in the tests that the volumetric stream amounts to q_M = 30 – 90 [dm³/h].

Various fabric barriers were placed in the model filter. Each barrier was tested by using the same model water-oil mixture: tap water – diesel oil. The fuel characteristics is shown in Table 1.

Table 1. Properties of the diesel oil

	Property	Diesel oil
1	density at 20°C [g/cm³]	0.8753
2	dynamic viscosity at 40°C [cP]	14.72
3	kinematic viscosity at 40°C [cSt]	17.08
4	max . water content [%]	0.10
5	flash point [°C]	101

2.1 Unwoven fabric filtration barriers

Needled cloth (unwoven fabric) barriers used in the tests were made of hydrophobic polypropylene (PP, 4.1 dtex) and polyester (PES, 1.3; 3.6 dtex) fibres. Their characteristics are given in Table 2. The multilayer FOW fabric features a layer of thin glass fibres with a diameter of 4 µm. To prevent fabric deformations when the filter is in operation the fabric was strengthened at the manufacturing stage with ET cloth made of PES fibres (elana-2dtex). The needled cloths were then thermally treated to thicken their structure. Characteristic properties of the tested unwoven fabrics are their high strength, low flow resistance, uniform distribution of fibres and pores across the whole volume of the cloth. Besides, these multilayered fabrics can be made with different degrees of layer packing.

Table 2. Characteristics of unwoven fabric used for tests of filtration barriers

Lp.	Parametr	Fabric symbol J/PP/1449	FOW	FINET-POP 2
1	Type of filtration barrier	PP thermally thickened	1. PES (3,6 dtex) 2. Glass 1–2 μm 3. PP(3,6–5,4 dtex) 4. ET fabric	PP – thermally thickened
2	G.S.M.	400 ± 50 g/m^2	650 ± 50 g/m^2 (1030 g/m^2)	300 g/m^2±10%
3	Longitudinal tearing strength	1400 N/5 cm	420 N/5 cm (500 N/5 cm)	350 N/5 cm
4	Transverse tearing strength	1400 N/5 cm	1700 N/5 cm (2000 n/5 cm)	500 N/5 cm
5	Longitudinal elongation	10 ± 10%	163% (170%)	60%
6	Transverse elongation	10 ± 10%	80% (75%)	40%
7	Permeability at 200 Pa [l/m^{2}s]	250 ± 50	150 (200)	200
8	Working temperature	max. 90°C	max. 90°C	max. 90°C
9	Porosity [%]	39	92	44
10	Mean fiber diameter [μm]	20,2	14,6	20,26
11	Mean pore diameter [μm]	31,2	49,6	28,12
12	Internal surface area coefficient [m^{-1}]	$8,2\ 10^{-4}$	$2,18\ 10^{-4}$	$7,55\ 10^{-4}$

2.2 The test stand

The tests were carried out on an upgraded test stand adjusted to the examination of coalescence barriers. The schematic diagram of the test stand is shown in Figure 3. The tank (7) of 500 dm^3 capacity was fed with tap water. The piston pump (4) with a controllable capacity ranging from 0 to 350 dm^3/h pumped water from the lower part of the tank and passed it to the model oil separator (1) with a volume of 12 dm^3. An oil feeding pump (3) was fitted on the discharge side of the pump before a rotameter (6). Oil was proportioned by a *Dyspenser 338* type metering pump (capacity $q_v = 0 - 11.5$ dm^3/h with a mean accuracy of ±1 mm^3/cycle), that drew oil from the tank (2). The volumetric stream of the mixture was measured with the rotameter. The pressure difference on the model filter was measured with pressure gauges (8) and (9).

The oil content after the oil separator was measured with a HORIBA OCMA-220 analyzer with an accuracy of ±0.5 % of oil content in a sample.

The system was fed with 1000, 2500, 5000 and 10000 ppm (1%) diesel oil by changing the volumetric stream of the emulsion from 30 to 90 dm^3/h. The changes of temperature, pressure drop on the barrier, oil concentration in the purified mixture (at the outlet of oil separator) were recorded. The oil concentration was determined using a Horiba OCMA 220 analyzer.

Besides, the investigation included the distribution of oil particle size in the solution after filtration through the oil separator barrier, using an instrument made by MALVERN INSTRUMENTS Ltd., meeting the ISO 13320/01 standard. The method of laser diffraction is employed to find the volumetric and quantitative distributions in the 0.01–2000 μm range.

3 AN ANALYSIS OF OIL-IN-WATER EMULSION FLOW THROUGH AN UNWOVEN FABRIC BARRIER

The flow of emulsion through a filtration bed with a multilayered unwoven fabric barrier FOW is shown in Fig. 4. The flow process is characterized by the coalescence of oil droplets from oil-in-water emulsion that takes place inside the barrier pores. Three areas can be identified that differ in the degree of oil dispergation.

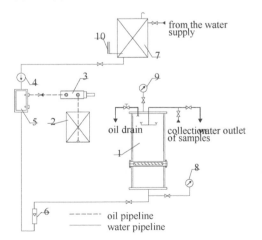

oil pipeline
water pipeline

Fig. 3. Test stand diagram; 1 – model filter with the tested barrier, 2 – oil tank, 3 – oil feeding pump, 4 – variable capacity piston pump, 5 – metering system, 6 – rotameter, 7 – water tank, 8, 9 – pressure gauges 0 – 0.1 MPa, 10 – thermometer

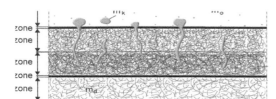

zone
zone
zone
zone
zone
m_d

Fig. 4. Emulsion flow through the unwoven fabric bed and the oil coalescence; md – oil (emulsion) mass before the barrier, mz – mass of oil stopped on the barrier, mk – mass of oil droplets released from the barrier; mo – mass of oil (emulsion) flowing through the barrier [4]

We observed that in external layer I the flowing emulsion does not visibly affect the system due to larger pores. The oil dispergation within this layer is the same as outside the barrier. In layer II droplets settle on the bed and on oil previously stopped on the packing structure. The diameter of droplets increases while they slowly move towards the bed centre. Large drops coalesce with small droplets and with those that already got bigger. As oil comes closer to the boundary of layer II, capillaries in the bed get filled up by drops contacting each other, but oil still has a discontinuous form. In layer III there is a characteristic flow of consistent oil stream through free space in the bed.

Layer I is only a few milimetres thin. The boundary between layers II and III, depending on a number of factors, shifts either in the direction of oil flow or in the opposite direction. The above description of emulsion flow through the bed is based on the assumption that the whole oil phase undergoes coalescence and leaves the bed in the form of drops much larger than those found in the emulsion.

One conclusion that might be drawn from the above is that pure water flows in capillaries of layer III not occupied by oil. However, this is not so, as part of the oil does not coalesce. Consequently, it will flow out of the bed together with the water phase in the form of very diluted emulsion. After coalescence, the droplet size increases. When a droplet is large enough, it will be pushed away from the fabric as an effect of resistance force acting against the flowing liquid. Once out of the bed, large drops can be easily removed by gravitational separation.

3.1 The efficiency of oil separation by coalescence

If we start from the oil mass balance on the barrier and assume that the oil concentration at the inlet is constant (Fig. 3):

$$m_d = m_z + m_k + m_o \qquad (8)$$

where: m_d – mass of dispersed phase before the barrier, m_z – mass of dispersed phase stopped on the barrier, m_k – mass of dispersed phase released as

droplets from the barrier, m_o – mass of dispersed phase remaining in the stream after the barrier.

Then the efficiency of filtration can be adopted as:

$$\eta = \frac{m_k}{m_d} = 1 - \frac{m_o}{m_d} \qquad (9)$$

When oil concentration is introduced [2, 6]:

$$\eta = 1 - \frac{C_o}{C_d} = 1 - e^{-\frac{4\beta\eta_i l}{d_w(1-\beta)}} \qquad (10)$$

where: C_o – oil concentration at the barrier inlet [ppm], C_d – oil concentration at the barrier outlet [ppm], $\beta = V_w/V_c$ – densitz coefficient of filter packing structure, V_w – fiber volume in the filter [m³], V_c – filter volume [m³], η_i – efficiency of filtration on the surface area of a single fibre, d_w – fiber diameter [m].

The efficiency of coalescence filtration (capturing oil droplets on a single isolated fibre) [6] is defined as the ratio of cross-section of part of the stream from which oil droplets will be separated to the cross-section area of a fiber set at a right angle to the flow direction (Fig. 5):

$$\eta_i = \frac{l \cdot 2l_p}{l \cdot 2r_w} \qquad (11)$$

where: η_i – filtration efficiency of a single fibre, l – fiber length [m], l_p – width of the cross-section stream part from which all oil droplets will be separated [m], r_w – fiber radius [m].

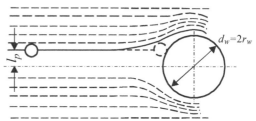

Fig. 5. The efficiency of filtration of a single fiber

The filter packing density equals:

$$\beta = Vw/Vc \qquad (12)$$

(for denotations see formula 10), and the flow speed in filter pores $\overline{w}$:

$$\overline{w} = \frac{W}{1-\beta} \qquad (13)$$

If the length of parallel cylindrical fibres in a filter volume unit is:

$$L_w = \frac{\beta}{\pi r_w^2} \qquad (14)$$

then the amount of mixture water and oil filtered by a filter with a thickness db:

$$Q_l = 2l_p \overline{w} L_w \, \mathrm{d}b = 2l_p \overline{w} \frac{\beta}{\pi \, r_w^2} \mathrm{d}b \tag{15}$$

A change of oil particle concentration in the flowing emulsion can be expressed as follows:

$$-\frac{\mathrm{d}c}{c} = \frac{\beta}{\pi \, r_w^2} \frac{2l_p \overline{w} \, \mathrm{d}b}{W} = \frac{\beta}{\pi \, r_w^2} \frac{2l_p \, \mathrm{d}b}{1-\beta} \tag{16}$$

After integration, we obtain:

$$\frac{c}{c_0} = \exp\left[\frac{-2\beta \eta_i b}{\pi \, r_w (1-\beta)}\right] \tag{17}$$

For the adopted layer of cylindrical parallel fibres and for monodispersed oil, we can express the efficiency of filtration in the form of the filtration equation (10):

$$\eta = \frac{c_0 - c}{c_0} = 1 - \exp(-\alpha_f) \tag{18}$$

where: η – efficiency of filtration (purification) of the fibrous layer, $\alpha_f = \dfrac{2\beta b \eta_i}{\pi \, r_w (1-\beta)}$ – coefficient of filtration of the fibrous layer.

The separation of oil particles from the flowing mixture of oil and water in the fabric layer occurs when:
– oil contacts the fiber surface,
– there exist forces holding oil particles in fiber surfaces.

Oil particles get in contact with fiber surface when their tracks come across the fiber surface. Several forces can be identified that cause oil particle tracks to diverge from the direction of water flow:
– inertia,
– water viscosity,
– diffusion,
– electrostatic,
– others: magnetic, terrestial gravity, thermoforesis etc.

Forces due to water viscosity tend to keep oil particles along the stream direction, while the other forces counteract it – make oil particles deviate from the stream line.

The efficiency of purification (formula 18) determines the quality of the filtration process. The accuracy of filtration is characterized by the mean size of contaminant particles stopped by the filter.

3.2 Results of oil separation efficiency tests and volumetric and quantitative distribution of oil droplets in unwoven fabrics

Figures 5, 6, 7 and 8 include the results of oil separation efficiency $C_o = f(q_v)$ and $\eta_o = f(q_v)$ obtained from two types of barriers FOW and POP2.

The values of oil concentration after the fibrous barriers do not exceed 50 ppm (FOW) and 16 ppm (FINET POP2) at 1% oil concentration before the barrier, while at 0.1% oil concentration, the figures at the outlet ranged from 4 to 20 ppm (Figs. 6–9).

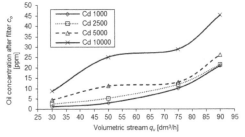

Fig. 6. The efficiency of oil separation by a FOW barrier FOW; $c_o = f(q_v)$

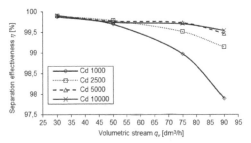

Fig. 7. The efficiency of oil separation by a FOW barrier; $\eta = f(q_v)$

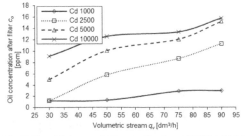

Fig. 8. The efficiency of oil separation by a FINET POP2 barrier FINET POP2; $c_o = f(q_v)$

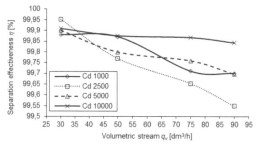

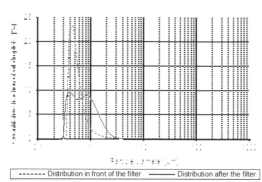

Fig. 9. The efficiency of oil separation by a FINET POP2 barrier; $\eta = f(q_v)$

The analysis of oil separation efficiency obtained on coalescence fibrous barriers of FOW and POP2 type leads to an observation that the following factors affect filter performance [1,3,4,7,8,22,23]: filtration bed properties (of the fabric itself, form of structure packing), process parameters (Δp in the bed, R_e), properties of oily water (c_d, ρ_o), volumetric distribution of oil particles, presence of SPC, mechanical contaminants.

The distributions of oil-in-water emulsion particles measured at the fibrous barriers FINET POP2 and FOW are given in Figures 10,11,12.

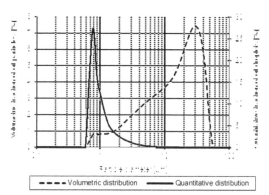

Fig. 10. Distributions of oil particles after the filter, FINET POP 2 barrier

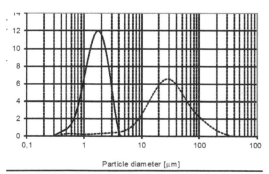

Fig. 11. Comparison of the volumetric distribution of oil particles on the FOW barrier

Fig. 12. Comparison of the quantitative distribution of oil particles on the FOW barrier

4 CONCLUSIONS

The tests show that FOW and FINET POP 2 barriers can yield satisfactory efficiency of oil separation (Figs. 5, 6, 8) at $q_v = 30$ dm³/h with oil concentration at the inlet 1000÷10 000 ppm. Higher concentrations result in correspondingly higher oil-in-water content after the barrier, ranging from 23.8–45.5 ppm, more than the standard requirement 15 ppm.

It follows from the above observations that unwoven fabric barriers can perform well as diesel oil separators when volumetric stream of the flow does not exceed, respectively 50 and 75 dm³/h, which guarantees relatively efficient work of the oil separation device.

Another advantage observed in the tests was a low flow resistance (over the entire range of tests), which translated into pressure drop of not more than 0.03 MPa.

An analysis of the distributions of oil droplet size for two barriers POP 2 and FOW (Figs. 9, 10 and 11) has shown that in terms of quantity particles with a diameter of 0.87μm make up 7.5%, while those with a 30 μm diameter occupy 27% of the total oil volume (Fig. 9).

The measured efficiency of 99.7% for both barrier types is comparable and does not differ significantly from other porous barriers [1,4,5,7,8,13,15, 22,23].

A filtration unit fitted with such barrier is capable of efficiently separating oil even at very high concentrations of oil (diesel oil) at the inlet, as confirmed by the lab tests.

As unwoven fabric is cheap, fabric filters in oil separators do not have to be reconditioned.

LITERATURE

[1] J. Chien, E. Glooyna, R. Scheeliter, Device for evaluating coalescence of oil emulsions, Journal of the environmentalenta lengieering division, April 1977.

[2] Ciborowski, Podstawy inżynierii chemicznej, WNT, War-
     szawa 1970.
[3] J. Grudziński, Badanie wpływu chemicznych środków
     myjących stosowanych na statkach na skuteczność odole-
     jania, Studia Nr 21, WSM, Szczecin 1994.
[4] J. Grudziński, Filtry koalescencyjne z przegrodą włókni-
     nową do odolejania wód zęzowych i balastowych na stat-
     kach, Praca doktorska, Raport 61/85, Politechnika Wro-
     cławska, Wrocław 1985.
[5] M. Gryta, K. Karakulski, A. Morawski, Purification of
     oily wastewater by hybrid UF/MD, Wat.Res., 35(15)
     (2001) 3665–3669.
[6] Hebda i inni, Filtracja oleju, paliwa, powietrza w silni-
     kach, WNT, Warszawa 1979.
[7] J. Hupka, Podstawy wykorzystania koalescencji w złożu
     w technologii odolejania wód odpadowych, Rozprawa
     habilitacyjna, Politechnika Gdańska, Wydział Chemii,
     Gdańsk 1988.
[8] P. Jeater, E. Rushton, G.A. Davies, Coalescence in fiber
     beds, Proceding of the Filtration Society Birminham Fil-
     tration and separation, IV 1980.
[9] G.V. Jeffreys, I.L. Hauskley, AIChEJ., 11, 413, (1965).
[10] W.M. Langdon, D.T. Wasan, Separation of organic dis-
     persions from aqueous medium by fibrous bed coales-
     cence, Recent Developments in Separation Science. Vol.
     V. 159–182, CRC Press Inc., West Palm Beach, Florida,
     1979.

[11] M. Malaczyński, Technika ochrony przed zanieczyszcze-
     niami ze statków, Wydawnictwo Morskie, Gdańsk 1979.
[12] Międzynarodowa Konwencja o Zapobieganiu Zanie-
     czyszczaniu morza przez Statki (Marpol 73/78), PRS,
     Gdańsk 2007.
[13] Moraczewski, Materiały filtracyjne do filtracji powietrza i
     wody, Ochrona Powietrza nr 3, 1986.
[14] Nowa Konwencja Helsinka, HELCOM, Dz.U. nr 28 poz.
     346, z dnia 14.04.2000.
[15] Pat. PL 1400038 1987.
[16] Pat. RFN 2609847.
[17] Pat. RP 130413 1987.
[18] Pat. USA 3969719.
[19] Prospekty firm: FRAM, VOKES, VAMAG Werke
     GmbH.
[20] Prospekty firm: KORDES, AWAS, Passavant, AWK
     Unikom System.
[21] Prospekty firmy MITPOL Sp. z o.o., Łódź 2003.
[22] L.A. Spelman, S.L. Goren, Experiments in coalescence by
     flow through fibrons mats, Ind. Eng. Chem. Fundus. Vol
     1, no. 1, 1972.
[23] L.A. Spielman, Su. Yeang-Po, Coalescence of oil in water
     suspensions by flow through porous media, Ind. Eng.
     Chem. Fundam. Vol. 16. No. 2, 1977.
[24] Wiewióra, Ochrona środowiska w eksploatacji statków,
     Fundacja Rozwoju WSM w Szczecinie, Szczecin 2005.
[25] Wiewióra, Z. Wesołek, J. Puchalski, Ropa naftowa w
     transporcie morskim, Trademar, Gdynia, 2004.

Ship Propulsion and Fuel Efficiency

16. Optimization of Hybrid Propulsion Systems

E. Sciberras & A. Grech
MI-SE@MALTA, MARSEC-XL Foundation, Senglea, Malta

ABSTRACT: Powertrain hybridization permits the benefits of more than one power source to be integrated and exploited for a beneficial effect on an objective, such as reduction of fuel consumption or emissions. Due to their operating profiles however, marine hybrid vessels do not exhibit much opportunity for free energy recuperation. Fuel savings can be realized by bettering component operating points, yet this requires correct sizing matched to the expected usage. In this paper, a multi-objective genetic algorithm is used to optimally size propulsion components in order to minimize fuel consumption as well as installation weight for a hybrid motoryacht operating on a day cruise scenario.

1 INTRODUCTION

Hybrid vehicles are now well established on land as a viable mode of greener transportation. The use of multiple energy sources and converters permits their individual benefits to be better utilized, by exploiting the inherent disparity between peak and average power demands (Schofield et al. 2005).

At sea, powertrain hybridization would equally permit the power demand to be met more effectively than by a single source. Yet marine hybrids are still not as popular as on land. 'Conventional' hybrids on marine vessels include diesel-electric systems, popular on passenger vessels, as well as CODLAG systems found on naval vessels. Such configurations of parallel electric and mechanical propulsors permit better efficiencies at part-loading and low speeds, due to the different sources being better suited for different loadings (Woud & Stapersma 2002). These hybrids however, differ from automotive ones in that they lack an Energy Storage System (ESS), typically in the form of chemical batteries.

The inclusion of an ESS would permit the loading of the prime movers to be optimized for greater periods of time, by using the ESS as a load bank during periods of low propulsion demand. Compared with automotive vehicles however, propulsive power demands for marine vessels are significantly larger; hence, by proportional scaling, the corresponding ESS would be excessively large, with an associated cost and weight factor.

The major shortcoming for marine hybrids stems from a lack of significant regenerative capability. A significant proportion of the energy efficiency for automotive hybrids comes from regenerative braking (Lukic et al. 2008). This permits energy which would otherwise be dissipated as heat at the brakes to be recovered to recharge the ESS. However, the lack of stop signs and traffic lights at sea much reduces the scope for energy recovery from deceleration. This is most apparent when comparing typical demand profiles between the New European Driving Cycle (a European standardized profile) representing a typical automotive suburban commute, and a typical day cruise for a marine vessel (Figure 1).

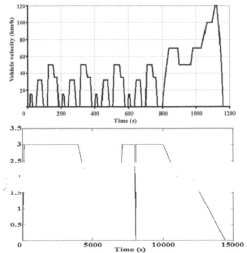

Figure 1. Comparison of automotive (top) and marine vessel (bottom) propulsion timelines (Barabino et al. 2009).

Fuel savings in the case of a marine hybrid are hence possible through correct sizing of components, such that overall operating points are improved over a particular scenario. Defining the fuel consumption for a scenario therefore requires a model for the hybrid system, which takes the scenario power demand as its input.

2 MODELLING

The optimal sizing of the hybrid system is simply the tip of the iceberg in the hybrid design process. Essential for the correct sizing is the demand profile, on whose realism the accuracy of the sizing will depend.

The power demand timeline for a marine hybrid consists of two parts, namely the propulsion demand and the hotel load demand. Also differing from automotive hybrids is a more significant hotel load, since motoryachts generally need to support onboard users for longer periods.

In determining the fuel consumption, consideration must be given to the interaction between prime mover, ESS and power demands. This requires a complete model of the hybrid system which considers all the power flows between the various components.

This model was built in Simulink, since no simulation tool was readily available for marine vessels. A sixty foot motoryacht was considered, for which trials data was available. A parallel hybrid configuration was proposed for this existing boat, by the addition of a battery bank and an electric motor/generator coupled to each diesel engine by a gearbox. The separate diesel generator could then be omitted by supplying the hotel load from the main battery bank and main engines.

From the trials data, the propulsive power demands were input as a Look-Up Table (LUT), returning the demanded power for the demanded speed. This converts the speed demand timeline to a power timeline. The diesel engine is modeled similarly, by converting the engine's performance chart into a two-dimensional LUT, taking engine speed and power as inputs, and returning the instantaneous specific fuel consumption (SFC). The cumulative fuel consumption is then the integral of the SFC values. The electric machine is modeled by its performance characteristic, with the power splitting and sharing being determined by a central control logic.

This steady-state modeling is valid since the quantities of interest (power flows and operating points) are required over a long period of time. Hence, transient response is not of particular interest for scenario fuel consumption determination. The batteries are modeled using Simulink's built-in battery model. This provides a model for Lithium-ion, Lead-acid and Nickel Metal Hydride batteries.

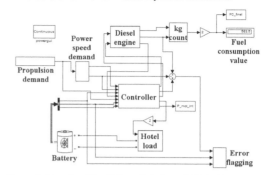

Figure 2. Complete Simulink model of parallel hybrid setup.

The central control logic controls the power demanded from the electrical machine and/or diesel engine, depending on the propulsion and hotel loadings, as well as the current operating point of the components. Critical above all is the batteries' state of charge, which is to be maintained within certain limits.

3 OPTIMIZATION

Hybrid vehicle design is generally approached from a satisfaction of specification. In a parallel automotive hybrid, an internal combustion engine (ICE) is sized to cater for the cruising speed demand, such that maximum speed on top gear is capable of being maintained. The low-speed side of the demand in turn influences the electric motor sizing. Together with the transmission system in use, this determines the acceleration capabilities of the vehicle. As a first-order design, the ICE can be assumed to cater for the steady-state rolling and air resistances, such that the electric drive is sized to completely meet the acceleration specification. The size of this motor can then be lowered by examining the power demanded for acceleration taking also into account the power provided by the ICE at low speeds (Ehsani et al. 2010).

For the ESS, the power requirement is selected to be greater than the motor's power rating to take into account conversion inefficiencies. The energy requirement is then dependent on the driving pattern to be catered for, and hence its regeneration potential. Taking into account the inefficiencies associated with the process and the desired initial and end capacities, then the stored energy requirement can be calculated. This design is then followed by simulation, when values such as fuel consumption can be calculated. Iterative design can then be performed in order to improve any aspect of the system (Ehsani et al. 2010).

Yet with such a design for satisfaction of specification, attributes such as fuel consumption, emissions and system weight are secondary values over which the designer has no direct control. Intuitive design, and experience help to direct the design and improve these parameters, however, the design does not address these parameters as an objective.

Optimization is a process whereby an objective is addressed directly and an extreme value (either maximum or minimum) located. This permits objectives to be aimed for and designed for, rather than following as a secondary consequence from design.

Classical optimization techniques would involve the use of mathematical tools such as the Newton-Raphson or steepest descent methods. These however require a mathematical equation for the problem description, something which can't be done to quantify the fuel consumption over a scenario. Furthermore, these methods all consider continuous and linear functions. When considering discrete component availability, classical optimization techniques fail for this problem.

Genetic algorithms take a cue from nature as the ultimate optimizer. Without requiring in depth knowledge of the problem at hand, genetic algorithms operate directly on a descriptor of the problem, treating the underlying function as a black box, requiring only the returned value. This robust approach based on simulation is therefore highly adept at optimizing hybrid vehicles, evidenced by works such as (Desai & Williamson 2009), (Jain et al. 2009) and (Hasanzadeh et al. 2005).

All the possible combinations of components making a hybrid setup represent the search space, from which the optimal configuration is chosen. In keeping with the genetic analogy, the descriptor for the component configuration is termed a *chromosome*. Corresponding to each chromosome in the *search space* is a *solution* in the *objective space*. This maps the chromosome to the objective value of interest such as fuel consumption.

The mapping from search to objective space is performed by the fitness function. Optimization is therefore performed on the solutions in the objective space, returning the fittest chromosome as the implementation to be selected.

Compared to classical methods, genetic algorithms are global routines, capable of locating population optima, rather than local ones. This is done without knowledge of any auxiliary parameters such as derivatives of the function, enabling genetic algorithms to be a robust method of global optimization.

Operating solely on the chromosome representation, the search for optima revolves around three operators. Considering a population of chromosomes, the *selection* operator identifies the fitter chromosomes to be used to generate the next generation. The next generation comes about by *reproduction*, whereby the previously selected chromosomes are used to form a new chromosome, termed the offspring. This represents the search through the search space and is responsible for locating the global optimum. Finally, the *mutation* operator provides an insurance against premature convergence by introducing a random variation to offspring to ensure that the search does not become stuck at a local optimum.

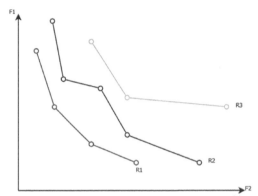

Figure 3. Three non-dominated ranks for bi-objective problem.

3.1 *Multi-objective optimization*

Despite the apparent straight-forwardness of optimization using genetic algorithms, optimization for a single objective does not reflect real-world practicalities. Locating an optimum with respect to a single objective would give an optimized solution, yet one which inherently ignores any other aspect of the problem. Referring to the problem of a hybrid motoryacht, optimizing for fuel consumption would result in a large battery capacity (to minimize engine operation and hence fuel consumption), yet come in at a large weight and cost.

Such a solution would be impractical from an application point of view, so a compromise must be found between locating an optimized solution from the consumption perspective, as well as the weight or installation point of view. Compromise should not imply substandard performance, but rather an addressing of differences.

A multi-objective optimization problem can trivially be converted to a single-objective one by means of a weighting vector, where multiple objectives are added up after being weighted to form a single metric. This however requires a priori knowledge of the demanded weighting. Results can therefore be biased since this decision is taken without any indication of results.

Basing the weighting after obtaining a set of results is possible by using the concept of non-domination and Pareto-ranking of solutions. Instead of delivering a single final solution, a set of optimized, compromise solutions is obtained, from which the final solution is chosen by the user using higher-level information. This higher-level information is experience-based and generally reflects non-technical influences, such as preference for particular components, or an inclination towards individual objectives. Though in effect this represents the use of a virtual weighting vector, the weighting values are applied to a set of results, thus the selection is based on actual solutions without postulating and introducing blind biases (Deb 2001).

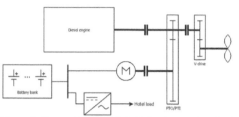

Figure 4. Proposed parallel hybrid implementation for hybrid motoryacht.

A very popular and efficient algorithm implementing a Pareto-based approach is the NSGA-II developed in (Deb 2002). The population is quickly sorted into ranks using the concept of non-domination, whereby a solution is said to be non-dominated with respect to another, if in going from one to the other, a certain sacrifice is demanded in one objective for a gain in the other, clearly illustrated as Figure 3. This shows a number of ranks, with R1 being the fittest rank. There is no benefit in choosing a solution from the lower ranks, but they can be used to search for new solutions, possibly giving better results.

Solutions in the first rank are the fittest, and this ranking value is used for selection purposes, as opposed to an explicit fitness value. This permits the comparison of solutions with multiple objectives. In order to further prioritize solutions for selection, a crowding metric is used to identify solutions lying in more isolated locations. This emphasizes a search in zones still unpopulated to enhance the global nature of the search.

4 IMPLEMENTATION

The model of the proposed parallel hybrid (Figure 4) was built in Simulink as outlined previously, with the genetic algorithm coded in Matlab.

The aim was to minimize both fuel consumption as well as installation weight, in order to determine the best compromise solution. The demand timelines are given as Figure 5 for both the propulsion as well as the hotel loads. The components to be optimized are the diesel engine, the electric motor/generator, the gearbox ratio and battery capacity as well as type. Optimization is also performed on the controller itself. This allows an even broader search space

and permits the exploration of different control strategies.

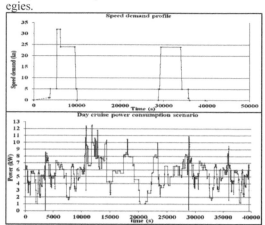

Figure 5. Propulsion (top) and hotel load demand timelines for sixty-foot motoryacht for day cruise scenario

4.1 The controller

The control strategy determines the points at which the vessel changes operating modes. For a parallel hybrid, four basic modes are identifiable, namely:

- Electric-only mode – all loads are supplied by the electric system from the batteries.
- Conventional mode – the diesel engines provide propulsion while the hotel load is supplied via inverter from the batteries.
- Assist mode – the electric motor connected to the batteries is used to assist the diesel engine during acceleration or high power demands, with their power added up at the gearbox.
- Charging mode – the diesel engine is run to provide propulsion and also supply the electric generator to recharge the batteries. Hotel load is supplied off the electric generator.

A speed and/or power level can be defined to control the changeover of modes, depending on the battery state of charge. Charging mode is enabled whenever the battery is discharged, while the other propulsion modes are only possible if the charge level is sufficient.

Operating the diesel engine at low power levels will result in high SFC values, in addition to suboptimal performance in terms of combustion, leading to higher wear and maintenance requirements. Thus, using electric propulsion for low demands is an obvious candidate for improving fuel consumption. However, raising the point to which electric propulsion is maintained necessitates increasing the battery size. Hence, the correct balance must be found. Likewise, the point at which assist mode is demanded can permit engine downsizing, but can lead to significantly longer charging times.

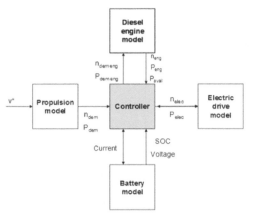

Figure 6. Power flows, with component set points decided by controller

The point at which assist is performed is a function of the diesel engine's loading, and hence the level of parallel operation demanded between motor and engine. Varying this level therefore allows the assist point to be optimized in order to determine the best load sharing. The changeover from electric-only to conventional mode is defined mainly by the electric motor's power and speed ratings, since electric operation is permitted only in this window.

Figure 6 illustrates the relation between the controller and the other simulated components. Based on the power demands and each components' current operating point, the controller outputs the desired setpoints for each component depending on its control strategy.

4.2 *Chromosome representation*

Based on these variables for optimization, the chromosome for searching through the search space was defined as consisting of the following elements:
- Diesel engine index
- Battery type
- Number of parallel batteries
- Electric motor rating
- Gearbox ratio
- Engine power sharing point
- Electric-only launch power

These all represent a particular hybrid setup from a database of components taken from manufacturer brochures. Thus every solution actually represents implementable setups. Real number representation is used, since this permits infinite database growth (without requiring chromosome modification as with binary coding) as well as avoiding Hamming cliffs which present an artificial hindrance to a gradual search (Deb 2001).

Using real numbers requires some modification to the standard algorithm, namely that a blending operator is used instead of explicit crossover. BLX-α was

implemented with an α-parameter of 0.5 to give the best balance between exploration and exploitation (Herrera et al. 1998).

4.3 *Crowding-distance metric*

The aims of multi-objective optimization are to identify the fittest possible set of compromise solutions, as well as explore the search space for a broader scope to this set. Deb proposes a crowded distance metric which identifies the biggest rectangle which can be fitted around a solution in the objective space (Deb et al. 2002). Yet this was found to give unsatisfactory results in this implementation, with limited final solution diversity. This is explained as being due to solutions having different chromosome makeup, yet giving similar solution values, thus decreasing an objective space metric's effectiveness.

This was further noted by Desai in (Desai & Williamson 2009) for a similar application. Desai's approach in the search space involved calculation of the Euclidean distance for between each point. This however is quite computationally intensive. The authors propose a novel *uniqueness* counter which counts the number of repetitions for each element for each chromosome in a population. Figure illustrates the functioning of this uniqueness counter on a sample population. This serves as an indicator as to how unique a solution actually is. Thus, during selection, in case of a tie between two solutions of equal rank, a more *unique* solution is preferred to ensure future diversity.

5 RESULTS

The algorithm was run for 100 generations in order to iterate towards the optimal rank of solutions. The equipment data was loaded from the component database, while the scenario hotel and propulsion timelines were obtained from previous work carried out within MI-SE@MALTA for a day cruise scenario for the 60-foot motoryacht under consideration (Grech 2009).

repetition counts

Figure 7. Uniqueness counter on sample population

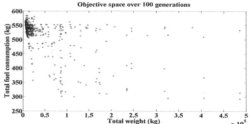

Figure 8. Solutions in objective space over 100 generations. Note convergence towards left hand side of space

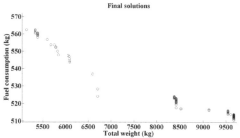

Figure 9. Final rank of optimized solutions

A population of size 200 was used, together with a mutation constant equal to the reciprocal of the chromosome length (Deb 2001). This gives a mutation rate proportional to the number of variables involved. A constraint of 10 tons is also introduced. This serves to focus the search below a total weight of 10 tons, representing a realistic figure which would otherwise involve a significant performance loss due to the added installation weight. It must be noted that as a first order model, the demand power is considered to be independent of loading, though in actual fact increased loading would increase power demand and correspondingly the fuel consumption.

The progression of the genetic algorithm is seen in Figure 8, where starting off from a random distribution in the objective (solution) space, the solutions increase in fitness by gradually migrating towards the left hand side of the objective space.

Figure 9 illustrates the final rank of optimized solutions. These are all rank 1, expected since an overall fitness improvement is demanded. Infeasible solutions (greater than 10 tons) are not illustrated in this figure. It is from this plot that the final solution is chosen by the user, coupled with further information obtained from examination of the solution chromosomes themselves.

A sample of the solution chromosomes is listed in Table 1. These chromosomes correspond to the solutions observed in Figure 9 in the objective space. All solutions utilize the same diesel engine as the conventional system (895kW rated power). This is understandable since the top speed requirement is not reduced, which demands around 800kW of propulsion power. Though electrical assistance is possible, the energy capacity required from the batteries would be excessive, resulting in a very heavy solution, and hence these solutions are dominated and discounted in early generations.

Also universally chosen was the option of having no gearbox connected to the electrical machine. Previous work (Sciberras & Norman 2010) without controller optimization had indicated a trend towards high speed machines coupled to a reduction gearbox. The controller optimization however now allows the motor's operating point to be variable and hence the additional weight of a gearbox can be avoided by locating a different launching power value.

The results clearly indicated the trend towards an energy dense solution. This involved Lithium-ion batteries and permanent magnet machines. Lithium-ion batteries offer the best specific energy capacity, essential for a marine hybrid where energy recuperation is largely absent. Though these involve significant cost compared to traditional lead batteries, their performance is highly superior (Lukic et al. 2008).

Likewise, permanent magnet machines offer greater power densities compared to conventional machines. This is due to the field excitation being provided by permanent magnets, removing the need for external excitation, and therefore greater efficiencies. This in turn implies a greater proportion of stored energy being converted to usable power. Permanent magnet machines are therefore more compact and lighter compared to their conventional cousins and are nowadays available off the shelf from several manufacturers. Permanent magnet machines also provide for more efficient generation capability.

The final setup choice is made by the user based on Figure 9 (visualizing the objective space) and Table 1 (illustrating the search space). Engineering experience and intuition now come into play, as well as reflecting preferences towards objectives. Aiding in the decision making, the user can visualize and examine the power flows for the selected solutions, such as Figure 10, by simulating a particular solution's behavior.

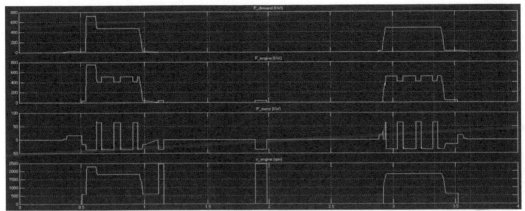

Figure 10. Component timelines for day cruise scenario. Chosen solution returns fuel consumption of 560.19 kg at a total weight of 5380kg.

Table 1. Selection of solution chromosomes after 100 generations. Repeated solutions have been omitted for clarity

Diesel engine rating	Motor rating	Motor speed rating	Total fuel consumption	Total weight	Battery capacity	Battery type	Diesel engine share point	Launch power
kW	kW	rpm	kg	kg	kWh		%	%
895	65	3000	562.40	5135.3	12.28	Li-ion	46	33
895	65	3000	562.40	5135.3	68.95	Li-ion	1	89
895	115	3000	562.20	5330.6	68.95	Li-ion	57	30
895	115	3000	562.22	5330.6	68.95	Li-ion	85	98
895	105	3000	558.99	5384.0	12.28	Li-ion	11	64
895	105	3000	560.19	5384.0	12.28	Li-ion	52	89
895	70	1800	553.70	5676.6	344.74	Li-ion	30	81
895	100	1800	553.69	5751.3	49.10	Li-ion	41	80
895	260	3000	552.51	5784.0	12.28	Li-ion	44	77
895	260	3000	552.12	5784.0	34.47	Li-ion	38	42
895	150	1800	549.97	5826.0	36.83	Li-ion	33	79
895	220	3000	547.86	5841.3	49.10	Li-ion	30	100
895	150	1800	547.55	6056.6	344.74	Li-ion	33	57
895	175	1800	546.92	6076.6	24.55	Li-ion	44	66
895	175	1800	546.92	6076.6	68.95	Li-ion	43	44
895	150	2400	523.55	8380.0	344.74	Li-ion	98	76
895	240	3600	521.52	8420.0	344.74	Li-ion	61	78
895	240	3600	522.88	8420.0	344.74	Li-ion	78	51
895	220	3000	520.31	8420.0	344.74	Li-ion	77	81
895	290	3600	516.94	8520.0	344.74	Li-ion	50	59
895	190	2400	517.44	8520.0	344.74	Li-ion	40	92
895	260	3000	516.86	9128.0	413.68	Li-ion	57	55
895	260	3000	516.32	9128.0	413.68	Li-ion	91	82
895	230	2400	516.09	9540.0	413.68	Li-ion	87	65
895	230	2400	515.99	9540.0	413.68	Li-ion	75	69
895	230	2400	514.00	9540.0	413.68	Li-ion	54	95
895	230	2400	514.56	9540.0	413.68	Li-ion	68	91
895	290	3000	513.92	9598.0	413.68	Li-ion	82	76
895	390	3600	512.62	9668.0	413.68	Li-ion	55	48

6 CONCLUSIONS

Objective design by simulation permits optimization of hybrid vehicles such that attributes such as fuel consumption can be aimed for and achieved by correct design. Classical optimization techniques are not able to successfully operate on complex models such as hybrid vehicles, hence genetic algorithms present a very powerful and robust way of arriving at optima by mimicking natural evolution.

A model was developed to calculate the fuel consumption of a hybrid motoryacht based on steady-state parameters. In turn, an optimization algorithm was developed to choose the best hybrid components as well as optimal controller values. This allows a hybrid vehicle to be virtually 'bred' from a computer.

Optimization is essential in marine hybrids, since the absence of regeneration implies that any savings must come about by improved component operating points. Intuitive design satisfies performance requirements, but does not guarantee fuel savings. This is emphasized by design by simulation, coupled with a robust optimization routine.

ACKNOWLEDGEMENTS

The work disclosed in this publication is based on work carried out at the Marine Institute for Software Engineering at Malta (MI-SE@MALTA) within the MARSEC-XL Foundation based in Senglea, Malta.

The research work disclosed in this publication is partially funded by the Strategic Educational Pathways Scholarship Scheme (Malta). The scholarship is part-financed by the European Union - European Social Fund.

REFERENCES

Barabino, G., Carpaneto, M., Comacchio, L., Marchesoni, M. & Novella, G. 2009. A new energy storage and conversion system for boat propulsion in protected marine areas. *Clean Electrical Power, 2009 International Conference on*: 363-369.

Deb, K. 2001. *Multi-Objective Optimization using Evolutionary Algorithms*. New York: Wiley.

Deb, K., Pratap, A., Agarwal, S. & Meyarivan, T. 2002. A fast and elitist multiobjective genetic algorithm: NSGA-II. *Evolutionary Computation, Transactions on*, vol. 6, no. 2: 182-197.

Desai, C. & Williamson, S.S. 2009. Optimal design of a parallel Hybrid Electric Vehicle using multi-objective genetic algorithms. *Vehicle Power and Propulsion Conference, 2009*: 871-876.

Ehsani, M., Gao, Y. & Emadi, A. 2010. *Modern electric, hybrid electric and fuel cell vehicles*. CRC Press.

Grech, A. 2009. *A day cruise scenario as an underlying foundation for a hybrid propulsion system optimization*. Tech. Rep. MARSEC09-432. Senglea: MARSEC-XL.

Hasanzadeh, A., Asaei, B. & Emadi, A. 2005. Optimum Design of Series Hybrid Electric Buses by Genetic Algorithm. *Industrial Electronics, 2005. ISIE 2005. Proceedings of the IEEE International Symposium on*, vol. 4: 1465-1470.

Herrera, F., Lozano, M. & Verdegay, J. 1998. Tackling real-coded genetic algorithms: operators and tools for behavioural analysis. *Artificial Intelligence Review*, vol. 12: 265-319.

Jain, M., Desai, C. & Williamson, S.S. 2009. Genetic algorithm based optimal powertrain component sizing and control strategy design for a fuel cell hybrid electric bus. *Vehicle Power and Propulsion Conference, 2009*: 980-985.

Lukic, S., Cao, J., Bansal, R., Rodriguez, F. & Emadi, A. 2008. Energy storage systems for automotive applications. *Industrial electronics, IEEE Transactions on*, vol. 55, no. 6: 2258-2267.

Schofield, N., Yap, H. & Bingham, C. 2005. Hybrid energy sources for electric and fuel cell vehicle propulsion. *Vehicle power and propulsion, 2005 IEEE Conference*: 522-529.

Sciberras, E. & Norman R. 2010. Sizing of hybrid propulsion systems. *Transactions in Evolutionary Computation*: Unpubl.

Woud, H.K. & Stapersma, D. 2002. *Design of propulsion and electric power generation systems*: 115-120. London: IMarEST.

17. Integrating Modular Hydrogen Fuel Cell Drives for Ship Propulsion: Prospectus and Challenges

P. Upadhyay, Y. Amani & R. Burke
SUNY Maritime College, Bronx, NY, USA

ABSTRACT: This paper proposes a new drive system for the ship propulsion. The drive power for propelling ship varies from few MW in a small cruise ship to hundreds of MW for large cargo ships. A typical cruise ship has a 6 MW drive whereas a cargo ship has 80 MW drive. Combustion drives are not sustainable and environment friendly. An idea of electric drive system using hydrogen fuel cell and necessary storage has been proposed. The hydrogen reformer develops hydrogen fuel cell using off-shore renewables like Wind, Wave and Solar power but the power handling capability of this fuel cell system (100 kW) restricts the application to the propulsion drives of several MW. The detail drive scheme describing; how multiple modular hydrogen fuel cell drives are integrated to develop variable power. The different options available for the propulsion system and factors affecting the choice are discussed in detail. Also, how such modular drives are helpful in controlling torque and power requirements is discussed. Replacement of electric drive reduces volume and weight of the ship and the available volume can be utilized for the storage and reform systems. The proposed paper will give a remarkable concept to overcome the challenges of utilizing hydrogen fuel cell to the larger scale and in future it can be extended to all other applications.

1 INTRODUCTION

The shipping industry has come a long way using combustion engines in transporting goods between continents, but in general the transportation sector accounts for a large fraction of air pollutant emissions. Health and environmental effects of air pollutants such as NOx, CO, VOCs, and particulates are leading to stricter tailpipe emissions regulations worldwide. [1, 2]. Virtually all transportation fuels today are derived from oil. Oil production is projected to peak worldwide within a decade and there no guarantee that oil will be enough for worldwide increase in consumption.

New frontiers have opened in the application of hydrogen as fuel for ships. A hybrid research ship [3], this research ship turns silent when scientists start recording whale songs. A 42-m long slick, hydrogen yacht with sufficient power was reported in [4].

This paper is a step toward resolving the integration of three main components required to design and discuss the concept of a large power hydrogen powered ship; the hydrogen fuel cell, the electronic drive and a uniquely designed motor. The uniqueness of the design is in integration 100kW, 250kW and the latest 4MW solid oxide fuel cells [5] to en-

ergize each pole of a sixty pole induction motor, assuming the drive system provides sixty Hz.

There is no doubt that there is much work to be done in order to establish a hydrogen based shipping industry, but the technology to initiate the transformation is in place.

The fact that the process and the technology is in place will require a resolution on how the new industry should be developed. The overall idea of a shipping economy based on hydrogen is discussed in [6,7].

2 HYDROGEN FUEL CELL

Energy can be stored in variety of forms, but clean energy production from natural sources such as wind, tidal, waves and sun are abundant, but need storage to be utilized when needed. The electric power produced by all means is used to split water into hydrogen and oxygen. Hydrogen could be stored in high pressure tanks, these tanks are getting smaller to a degree that it could be utilized in ship design. The DOE hydrogen program reports, tube trailer delivery capacity of 700kg by 2010 and 1,100kg by 2015 at 8300psi. Note that heating energy in hydrogen is 33.33kWh/kg, for methane it is 13.9-kWh/kg and for petroleum 12.4kWh/kg. To travel 400km, a modern combustion vehicle needs 24-kg of gas, 8-kg of hydrogen in hydrogen combustion engine and 4-kg of hydrogen in a fuel cell, electric drive vehicle.

A hydrogen fuel cell is a device that converts hydrogen to electricity. The device exhausts pure water and heat in the process of this transformation. Fuel cell technology has improved to a degree that many car companies are introducing new models of commercially available hydrogen cars. DOE projection of hydrogen fuel cell prices have been achieved, as this market expands the fuel cell prices will be much lower than DOE projection. Figure 1 depicts price trends for hydrogen fuel cell.

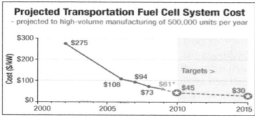

Fig. 1 Price Trend for Hydrogen Fuel Cell

To completely demonstrate the feasibility of the hydrogen based economy a model cargo ship, sixteen to twenty two mega watts should be constructed using hydrogen fuel cells with possibility of on board hydrogen reforming, production and storage.

To supplement hydrogen, we propose an aerodynamic cargo model ship with retrieving solar panels and small onboard wind turbines to produce power needed for in house hydrogen reformers.

Hydrogen production and dispensing is done by creating small scale hydrogen producing platforms, containing solar, wind, and wave energy conversion mechanisms. The electric energy produced is used to split water into hydrogen and oxygen. These platforms also house a fueling station and limited storage for hydrogen. A typical MERP- Marine Energy and Refueling Ports is depicted in fig.2.

Using technology developed for automobile hydrogen refueling stations and applying it to Marine Energy and Refueling Port (MERP) is feasible, however, more studies need to be conducted to relate the two systems. Here are a few proposed models:

1 The first model is suitable to coastal areas where a hydrogen pipeline is part of an existing infrastructure to bring energy to urban and coastal areas. The MERP's can be integrated into this system and could simply contain small storage and fueling stations.

2 The second model is based on a distributed, small scale local supply for hydrogen shipping. This model encourages complete reliance on renewable energy resources such as tidal, wind, solar and, where available, wave energy.

3 Designing and utilizing mobile hydrogen producing ships equipped with wind turbines and hydrogen reformers. These mobile vessels could catch high winds and produce hydrogen and could also serve as on the way mobile refueling stations.

Marine Energy & Refueling Ports are envisioned as non-intrusive, small islands attached to coasts or in off-coastal areas where maximum energy yields can be harvested from wind, wave and tidal currents. Such an MERP will house a compatible hydrogen reformer, low pressure hydrogen storage and a fueling station. In addition, these structures should house a vertical axis wind turbine, a vertical axis tidal turbine, photo voltaic panels and the appropriate electronics necessary for the control and conversion.

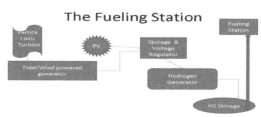

Fig. 2 Marine Energy & Refueling Port

Electric drives for ships are well developed and require little change when coupled to hydrogen fuel cells. The design proposed in this paper is an attempt to show how the technology developed for cars could be transferred to ship design.

3 ALL-ELECTRIC SHIP CONCEPT

Combustion drives will not be sustainable over the future and are not environmentally friendly. The electric drive system for the ship propulsion proposed in this paper has the advantages of [8]:
– Efficient and Improved life cycle cost
– High Power/volume and Power/weight ratio i.e. high payload of vessel
– Less propulsion noise and vibrations
– Ease of speed control
– Flexibility in thruster device locations

All-electric ships using fossil fuels are a present day reality. This concept leads to designs which can use on-board electric power for effective and efficient propulsion, while auxiliary systems usually powered by steam, hydraulic, or pneumatic energy are converted to electrical power. This "single bus" ship can thus allocate power as needed according to the mission profile of the vessel. Electric propulsion has been applied to different types of ships, such as cruise vessels, ferries, dynamically-positioned drilling vessels, thruster assisted moored floating production facilities, shuttle tankers, cable layers, pipe layers, icebreakers, supply vessels, and naval vessels. There are many different configurations available for the propulsion systems.

In conventional all-electric ship design configurations, sets of engine-generators produce electric power that is distributed for all auxiliary and main propulsion systems as shown in fig. 3. The system is approximately about as efficient as conventional non-electric drives, but the costs of generators, motors and static drives can make this solution expensive [9]. The operational benefits and the advantages of design flexibility justify such additional costs.

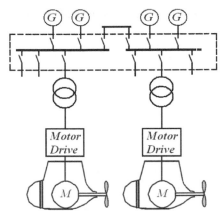

Fig.3 On-ship Power generation distribution and Propulsion Motor Controls

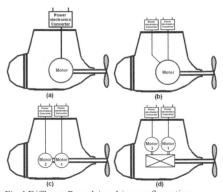

Fig.4 Different Propulsion drive configurations

In these configurations, the power developed by all generators is supplied to a common bus. Then the ship propeller is supplied power through transformers and converters as shown in fig.3.

Typically, for twin-screw ships, each propeller is controlled by a single motor drive as shown in fig. 4(a). Other common schemes used for E-Ship propulsion are shown in fig.4. A two winding motor with redundant converters is shown in fig. 4(b), in which the redundant winding is supplied through another power electronic converter. Fig.4 (c) shows Tandem motor with redundant converters, and fig.4

(d) shows a geared dual shaft propulsion drives in which, two motors are coupled to the propeller through a gear [10]. The application of these configurations depends on size and type of the ship.

4 HYDROGEN FUEL CELL BASED DRIVE

As discussed in section-1, the limitation of hydrogen fuel cell is to supply power at the level of 100 kW. For conventional diesel-generators and fuel based generators, one of the configurations discussed in section 3 can be used. Also, for boats and small ships, it is easier to handle propulsion power through one of the schemes shown in fig. 4. For cargo ships having power requirements much greater than 5 MW, it is difficult to address the challenge of making the ship all electric-ship utilizing hydrogen fuel cells. Two configurations are proposed in this paper namely (a) distributed modular generators, and (b) modular drive operation. Following section discusses each of the schemes in detail.

4.1 Distributed modular generators

A distributed modular generation configuration is shown in fig. 5. It is possible to mount the hydrogen fuel cell (FC) modules in a distributed form in the ship. An AC bus can run through the ship and all power is generated by fuel cells (FC) and controlled by a power electronic (PE) converter is supplied to this bus. A transformer provides the higher voltage required for the propulsion motor. The auxiliary supply can be met by either the similar fuel cell or using diesel-generator set. This configuration is advantageous when the fuel cells are located as distributed form. Control of each individual fuel cell power generator module is a challenge if they are located as distributed manner.

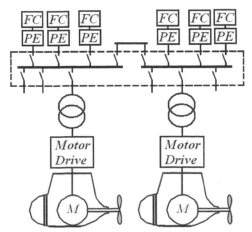

Fig. 5 Distributed Hydrogen Fuel Cell Power generations and Propulsion Motor Controls

4.2 Modular drive operation.

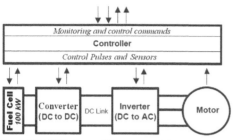

Fig. 6 100 kW Hydrogen FC power module

Another scheme is to place all hydrogen fuel cells centrally. Each of FC produces power in order of 100 kW. The PE module designed to work at low power rating of 100 kW, can control power flowing to the stator of the main motor as shown in fig. 6.

The sensors measure speed, position, load current and all necessary control parameters. The controller senses all these signals and send signals to centrally located control station for monitoring and control action. The controller also develops control pulses for the FCs, PE converters, and inverters. All these modules are synchronized with motor parameters. Due to low power and voltage ratings, the cost of this module is low as compared to developing large rating PE converters discussed in section 4.a, and a transformer is not required to boost the ac voltage.

A typical small ship may have a 6 MW drive whereas a large cargo ship has an 80 MW drive [5]. To control 6 MW propulsion power, sixty such modules are placed as shown in fig.5. All these modules are integrated through a centralized distributed control system. For higher power applications, Permanent Magnet Synchronous Motors (PM SMs) could be employed. The advantages of PMSMs include high efficiency, ease of control, and high torque/weight ratio. Due to reductions in the cost of rare-earth magnets, these machines have lower payback periods. Thus, designers may select among alternative configurations depending upon the size and type of application for ship propulsion.

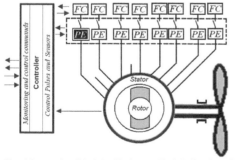

Fig. 6 Integrating Modular Hydrogen Fuel Cell Drives for ship Propulsion

5 SIZING EQUATION FOR PERMANENT MAGNET PROPULSION MOTOR

The output equation for the radial-flux PM BLDC motor is derived based on the expressions for the torque and back emf [11];

$$P_r = T\omega_m = \eta N_c E_{ph} I_{ph} \qquad (1)$$

$$T = \frac{P_r}{\omega_m} = \frac{\eta N_c E_{ph} I_{ph}}{\omega_m} \qquad (2)$$

By substituting the values of induced emf in the above equation, the torque is given as;

$$T = \frac{\eta N_c (N_m N_{spp} K_w B_g LD_{ro} n_s \omega_m / 2) I_{ph}}{\omega_m}$$
$$= \frac{\eta N_c N_m N_{spp} K_w B_g LD_{ro} I_s}{2} \qquad (3)$$

$$LD_{ro} = \frac{2T}{\eta N_c N_m N_{spp} K_w B_g I_s} \qquad (4)$$

The rated power output is the product of efficiency, phase voltage, phase current and the number of coil conducting simultaneously. The output is also given by the product of developed torque and the motor speed in rad/sec. Comparing the two and simplifying the equation the output equation for the radial-flux PM BLDC motor can be obtained. A specific slot loading I_s can be considered for the output equation. The LD_{ro} product depends on the torque developed by the motor, specific magnetic loading, specific slot loading, and the efficiency as shown below;

$$LD_{ro} = \frac{2}{\eta N_c N_m N_{spp} K_w B_g I_s} \left(\frac{P_r}{\omega_m} \right) \qquad (5)$$

Output equation relates the physical dimensions of the radial-flux PM BLDC motor with the power output, speed, assumed efficiency, number of phases conducting simultaneously, number of magnet poles, slots per pole per phase, winding factor, assumed magnetic loading and assumed electric loading [11].

For the rated power of 6 MW, 100 rpm, 1000 V per drive module, 60 poles, 3 slots/poles/phase and Average airgap flux density of 0.6 T, following overall machine dimensions are obtained;

Core Length of machine = 0.95 m
Rotor inner diameter = 0.95 m
Outer diameter of stator = 1.57 m
Efficiency of the machine = 0.96

Usually, the shaft diameter for 6 MW propeller is 0.9 m. This parameter matches with the rotor inner diameter. The machine can accommodate 60 coils for each power electronics module. An idea of electric drive system using hydrogen fuel cell and necessary storage has been proposed [2]. The hydrogen reformer develops Hydrogen fuel cell using offshore renewables like Wind, Wave and Solar power but the power handling capability of this fuel cell system (100 kW) restricts the application to the propulsion drives of several MW. The detail drive scheme describing; how multiple modular hydrogen fuel cell drives are integrated to develop variable power is shown in fig. 6.

6 ECONOMIC CONSIDERATIONS

The economic advantages of hydrogen-based ship propulsion remain uncertain at present, but may become more apparent as hydrogen production and consumption becomes widespread. The factors that must be considered for an economic analysis of hydrogen ship propulsion include:

- The weight, volume, and cost of shipboard hydrogen fuel storage compared to traditional storage of fuel oil.
- The weight, volume, and cost of electric generating equipment and main propulsion motors compared with traditional diesel or steam main propulsion machinery and associated ship's service generators.
- The cost of obtaining hydrogen fuel as compared to obtaining hydrocarbon fuels that will satisfy environmental requirements in the future, on an energy-equivalence basis.
- The cost of periodic maintenance of hydrogen-electric machinery compared to traditional marine power plants.

This assumes that the availability and reliability of hydrogen-electric machinery will be equivalent to traditional plants. This is a fair assumption with respect to the electrical machinery, but remains to be proven for fuels cells and related equipment. Also, one must assume that adequate supplies of hydrogen will be available.

Given a twenty year life for a ship, an incremental analysis of equivalent ships having alternative propulsion modes would rely upon a net present value expression such as:

$$NPV(\Delta Cost) = \Delta Cost_{MACH} + \Delta\ Cost_{FUELSYS}$$

$$+ (P|A,i\%,20)\ [\Delta AnnCost_{FUEL} + \Delta AnnCost_{MAINT}] \quad (6)$$

where the change in costs of machinery and fuel systems are capital expenditures in the present, and the sum of annual differences in the costs of fuel and maintenance are reduced to a single present value by

the application of the Series Present Worth Factor over the life of the ship at a cost of capital of i%.

For a cargo ship, the Minimum Required Freight rate (MRFR) is often used as a figure of merit is assessing a ship design. This is simply a ratio of the annualized cost of the acquisition and operation of the ship over the life of the ship, divided by the annual tonnage of cargo carried (i.e., the ATC), and the owner seeks to have a vessel with the minimal MRFR to be more competitive. Assuming that the cargo carrying capacity of a hydrogen-powered vessel is the same as a conventionally powered ship, the change in Minimum required Freight rate would be

$$\Delta MRFR = \frac{(A|P, i\%, 20)NPV(\Delta Cost)}{ATC} \quad (7)$$

A singular advantage that could accrue to a hydrogen-powered vessel could be a reduction in weight and volume of the machinery and fuel storage, which would allow for additional cargo to be carried in a ship of equivalent displacement.

7 CONCLUSIONS

Replacement of electric drive reduces volume and weight of the ship and the available volume can be utilized for the storage and reform systems. The proposed modular drive scheme will give a remarkable concept to overcome the challenges of utilizing hydrogen fuel cell to the larger scale and in future it can be extended to all other type of marine applications.

REFERENCES

[1] Jennifer Guevin-Global shipping pollution ain't pretty Green Tech. Feb.26, 2009
[2] John Vidal, Environment editor , guardian.co.uk
[3] Jerry Stanfield, "Moving Ahead" Yatch International ID Magazine, Winter 2010, pp 22-28
[4] Greg Trauthwein, "Back to the Arctic" Marine Technology, October 2009, pp 24-26.
[5] American Bureau of Shipping, ACTIVITIES, August 2010
[6] Yaqub M. Amani "Hydrogen Based Shipping Industry"AG11 Presentation, Pusan South Korea Oct. 2010
[7] Yaqub M. Amani "A pioneering Hydrogen Fueled Ship"
[8] The Institute of Marine Engineers, All Electric Ship, developing benefits for maritime applications, Conference Proceedings, UK, 29-30, September 1998
[9] The First Global Conference on Innovation in Marine Technology and the Future of Maritime Transportation
[10] Hackman, T. "Electric propulsion systems for ships" ABB Review, No.3 pp. 3-12, 1992
[11] Parag R. Upadhyay and K. R. Rajagopal, "FE Analysis and CAD of Radial-Flux Surface Mounted Permanent Magnet Brushless DC Motors", *IEEE Transactions on Magnetics* Vol. 41, No. 10, October 2005, pp 3952-3954.

Ship Propulsion and Fuel Efficiency
Miscellaneous Problems in Maritime Navigation, Transport and Shipping – Marine Navigation and Safety of Sea Transportation – Weintrit & Neumann (ed.)

18. Modelling of Power Management System on Ship by Using Petri Nets

M. Krčum, A. Gudelj & L. Žižić
University of Split, Faculty of Maritime Studies, Split, Croatia

ABSTRACT: Electrical power system of ship is consisted of power generators, consumers and distribution system. All parts of power system onboard, with or without its own controllers, are interconnected with control system – hardwired or field bus. That provides enormous possibilities for advanced overall control system functionality. The Power Management System (PMS) is a critical part of the control equipment in the ship. It is usually distributed on various control stations that can operate together and share information between each other or independently in case of special emergency situations in which ship have to operate. The equipment within the PMS includes the engines, generators, switchboards and controls along with the automation equipment that perform the calculation algorithms.

The design of future ship will require the development of new and increasingly sophisticated methods for the modeling and simulation of the complex systems that must be integrated in order to produce the total ship. Control architecture for power distribution systems has to be hierarchical, distributed and easy to adapt.

A complete logistic chain of this control architecture will be modeled by Colored Petri Net (CPN), which connects effective agents for autonomous control of complex distributed systems with agents for the control of power management systems.

After introducing the basic Petri Nets concepts, the cooperation and optimization process among the agents is modeled by the CPN Tools developed. The interdependency between these agents is mainly related through information and orders. The goal is to develop the model necessary to study the characteristics of complex systems and support the validity of the integrated control architecture. The results should show that the CPN model provides an effective simulation environment and enhances the power management system on ships.

1 INTRODUCTION

Ships need electric power for everything: from navigation lights to the automation system. Diesel – electric propulsion systems or standby electric propulsion systems, known as boosters, are being increasingly installed in many modern ships. There is hardly a piece of equipment or a system on board that will function without electric power.

Ships are autonomous systems. They have own power plant on board, and the demands placed on these power plants vary just as widely as the ships themselves. The problems are different: a lack of space, quiet running, low heat losses…. Environmental protection aspects are gaining importance. Above all, designers of on-board power systems concentrate on economy and reliability.

The most interesting issue is to develop optimum energy concepts for every type of ship, from time-proven standard solutions to individual design concepts fitted with all the necessary equipment: generators, transformers, frequency converters, rectifiers, inverters, uninterruptible power supplies, switchgear, power management systems, etc.

This paper discusses new techniques which aim to overcome the shortcomings of the protective system. These techniques are composed of advanced monitoring and control, automated failure location, automated intelligent system reconfiguration and restoration intelligent system reconfiguration and restoration, and self-optimizing under partial failure.

The use of shaft generators is rapidly becoming a popular method of producing electric power for the various electricity consumers on board ships. The new techniques will eliminate human mistakes, make intelligent reconfiguration decisions more quickly, and reduce the manpower required to perform the functions.

In this paper the environment of the agents is the power system. In a centralized Multi Agent System application, some agents works dominantly based on the preset global information of the power system or are responsible for coordinating or directing the functions of the other agents.

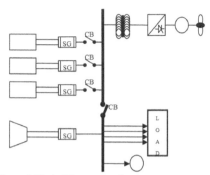

Figure 1. Typical Power electric system.

As a graphical and mathematical tool, Petri nets Dicesare, 1991., have been successfully used in communication protocols and automated manufacturing systems, in which they offer a flexibility to simulate discrete even systems. Therefore, we choose to use Petri nets as the modeling tool of the multi-agents systems for power system on ship. By extending PN models to the power system, we would like it to reflect the changes brought by the information between agents.

Figure 1 illustrates synchronous generators (power sources) transformers, power panels, bus transfer units and interconnecting cable used for delivering power to the loads.

Some equipment failures may lead to large over current conditions. Protective devices are used in electrical power systems to prevent or limit damage during abnormalities and to minimize their effect on the remainder of the system Krcum, 2005.

2 PETRI NETS

2.1 Basic theory about Petri Nets

PN is a graphical and mathematical modeling tool which can be used as a visual communication aid. Basically, PN is a bipartite graph consisting of two types of nodes, places and transitions, connected by arcs.

Petri net is a 6-tuple Murata, 1989.:

$$PN = \left(P, T, \mathbf{I}, \mathbf{O}, \mathbf{M}, \mathbf{m_0} \right) \tag{1}$$

where:

$P = \{ p_1, p_2, \dots p_m \}$ - set of places,

$T = \{ t_1, t_2, \dots, t_n \}$ - set of transitions,

$P \cap T = \varnothing$,

$\mathbf{I}_{(n,m)} : T \times P \to \{0,1\}$ - an input incidence matrix,

$\mathbf{O}_{(n,m)} : T \times P \to \{0,1\}$ - an output incidence matrix,

$\mathbf{M} : [\mathbf{I} : \mathbf{O}] \mapsto \{1, 2, 3, \dots \}$ - is a weight function,

$\mathbf{m_0}$ - initial marking.

Places and transitions $v \in P \cup T$ are calling nodes and denote states and events in the DES. A transition $t \in T$ is enabled at a marking $m(p)$ iff $\forall p \in \bullet t, m(p) \ge w\left(I(p,t) \right)$ ($\bullet t$ is a set of input places to transition t, and $w\left(I(p,t) \right)$ is weight of the arc between p, t. A transition t that meets the enabled condition is free to fire. When a transition t fires, all of its input places lose a number of tokens, and all of its output places gain a number of tokens. In a Petri net PN with m places and n transitions, the incidence matrix $\mathbf{W}$ is a $n \times m$ matrix where elements are $a_{ij} = w(t_j, p_i) - w(p_i, t_j)$. If all arcs in PN have weight equal to 1, it should be noted that

$$\mathbf{W} = \mathbf{O} - \mathbf{I}. \tag{2}$$

The matrices $\mathbf{I}$ (input matrix) and $\mathbf{O}$ (output matrix) provide a complete description of the structure of a Petri net. If there are no self loops, the structure may be described by $\mathbf{W}$ only. The incidence matrix allows an algebraic description of the evolution of the marking of a Petri net. The marking of Petri net changes from marking $m_k(p)$ to marking $m_{k+1}(p)$:

$$m_{k+1}(p) = m_k(p) + \mathbf{W}^\mathsf{T} \cdot \sigma. \tag{3}$$

where σ is a transition vector.

2.2 Informal Description of Colored Petri Nets

Colored Petri nets (CPNs) are based on extensions to normal Petri nets Kristensen, 1998., Jensen, 2007., CPN-Tools. CPNs extend the modeling capabilities of the traditional place-transition PT Petri net. Colored tokens can be defined from different types ranging from simple to complex. The token in a colored Petri net can encode a vast amount of information that determines transition firing. Places are associated with color sets. This specifies the tokens that the place can have. A transition can be programmed using special constructs and functions. Additional constructs can be used to enable or disable transition firing. Input and output arcs can have expressions and functions related to them.

For a transition to be enabled the input arcs expressions need to bind successfully with the tokens present in the input places and the transition guard. The tokens are placed in the respective output places.

CPNs being a class of higher order nets Garcia, 2008., Kristensen, 1998., Jensen, 2007., offer the advantage of having a memory state that is controlled via the tokens themselves. They can be used for fault diagnosis and investigation in control systems offering several advantages over traditional FSMs and PT PN Garcia, 2008.

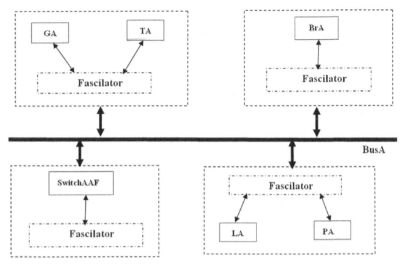

Figure 2. The overview of MAS for the PPS

3 AGENT TYPE

According to the electric component to which an agent is associated, the agents in the multi-agent system (MAS) can be classified into seven categories generator agent (GA), gas turbine agent (TA), load agent (LA), propelling system agent (PA), breaker agent (BrA), switches agent (SwitchA) and bus agent (BusA).

A generator agent can receive current information of the corresponding generator in the power system. This information consists of generation capacity, real/reactive power output, generation cost, fault alarm, etc.

The goal of agents LA and PA is to make sure that the corresponding load is supplied in the PPS. In this work, the propelling system has the highest priority number based on the importance of the load. The loads with higher priorities are more important than the load with the lower priorities, and need to be restored prior to the loads with lower priority.

Circuit breaker agent BrA interacts with the corresponding breaker in the power supplier layer. BrA receives the status of the corresponding breaker, and it can also send control signals to the corresponding breaker.

A bus agent monitors the current information of the corresponding bus in the power system. BusA updates the information by communicating with its neighboring agents and collects information from the corresponding bus in the power system and then sends the updated information to the neighboring agents.

Based on the information received from neighboring agents, the switch agent can reconfigure the power system by controlling the corresponding switches. Through the control, SwitchA can lead to connecting/disconnecting of a load shedding and load recovery.

The MAS overview for the Power Plant System (PPS) is shown in the Fig. 2 Facilitators are introduced into the GA, LA and BrA in the MAS to reduced the communication among the agents. GAF receives the information from GA and TA, updates the received information by interacting with the corresponding agent, and then forwards the updated information to the neighboring agents. If a fault is detected on the corresponding generator, the GAF sends information to the BusA to inform that the generator is fault and open the corresponding circuit breaker. If the corresponding supplier unit is currently disconnected from the system, and current power supply is less than the power demand in the system, the generator GFA sends a request to the BusA to close the circuit breaker, so that the generator can be reconnected to the PPS.

If LFA agent detects that the power supply is less than the power demand in the system, and the corresponding load is a load with low priority, LFA informs the neighboring agent to disconnect the load from the PPS. If the load has a fault on it, the corresponding load agent sends information to the neighboring breaker agent to inform that the load does not work properly. If the load is currently disconnected from the system, and the corresponding load agent detects sufficient power supply to supply the load, the load agent sends a request to the neighboring agent in order to initiate reconnection of the load into the system.

Based on the information received from neighboring agents, BrAF can reconfigure the power system by controlling the corresponding breaker. Through the control, the circuit breaker agent can lead to connecting/disconnecting of generators and a gas turbine.

4 CPNS MODEL

Since PN simulates all of the system states and all transition judgments by token passing in a quite straightforward manner, the graphical representation for a moderate system shows very complex configuration. In CPN a place node owns several colors to represent different states and base on the colors the judgment functions in a transition node checks the states of the incoming place nodes. The characteristics dramatically simplify the graphical representation of the traditional PN and improve the execution efficiency too.

So, by combining the MAS framework, algorithms and rules in CPNs, the CPNs model for the MAS is designed as the CPN tools (Version 2.2.0) shown in Fig. 4.

Places contain a set of markers called tokens. In contrast to low-level Petri nets (such as Place/Transition Nets), each of these tokens carries a data value, which belongs to a given type. As an example, place GA has one token in the initial state. The token value belongs to the type *status* and represents states of diesels generators. The first element in the pair is the pair $(dg(1),e)$ which denotes that the firs generator is energized, while the second is referred to the second diesel generator. The detail description of all colors is shown on the Fig. 3. The description of all the places and transitions are illustrated in Table 1 and Table 2 respectively.

Table 1. The Meaning of places in CPN.

Places	Description
PA	Propulsion Agent
LA	Load Agent
LAF	Load Facilities Agent
GA	Diesel engine agent
GAF	Generator Facilities Agent
TA	Turbine Agent
BrA	Breaker Agent
BrF	Breaker Facilities Agent
BusA	Bus Agent
PDP	Power Demand Plan (sum of power demand from load and propeller)
PA	Propeller Agent
LA	Load Agent
SwitchA	Switch Agent (for connecting/disconnecting loads)
GSP	Global Supply Plan. This place saves all initial supply plans submitted by power supplier agents

Table 2. The transitions and their description

Transitions	Description
BC	The PDP inform all the LAF that new organization is required.
Des	The Main Facilitator makes the global supply plan about loads, suppliers, breakers and switchers. It communicates to the agents and tells them the corresponding entities on their new task and announces that the new organization begins.
KDB	The agents learn from PDP plan and save the rules to the knowledge database.
LDF	Load Demand Forecast. Forecast the load demand and send to PDP.
re, reset	Breaker (switch) states are reset..
t1, t2	Diesel generators facilitator / Gas turbine facilitator make a local power supply plan, according to agent's plan saved in the knowledge database

```
▼Declarations
  ▼colset I=int;
  ▼colset B=int;
  ▼colset G=index dg with 1..2;
  ▼colset st1=with e|d|f;
  ▼colset st2=product st1*st1;
  ▼colset tp=with p;
  ▼colset supp=product G*st1;
  ▼colset tsup=product tp*st1;
  ▼colset BrL=with bl|bp;
  ▼colset status=product supp*supp;
  ▼colset S1=product status*tsup;
  ▼colset st3=product st1*st2;
  ▼colset DGe= product G*I;
  ▼colset DGe2=product G*I;
  ▼colset PR=product DGe*DGe2;
  ▼colset Br=product B*B*B;
  ▼colset S2=product S1*Br*st3*Br;
  ▼colset st4=with st1;
  ▼fun sw()=discrete(0,1);
  ▼fun SwS(swt:Br):Br=(sw(),sw(),sw());
  ▼fun new(l1:st2):st2=(st1.ran(),st1.ran());
  ▼var i:I;
  ▼var x2:st1;
  ▼var pdp:st1;
```
Figure 3. The color settings and variable declarations

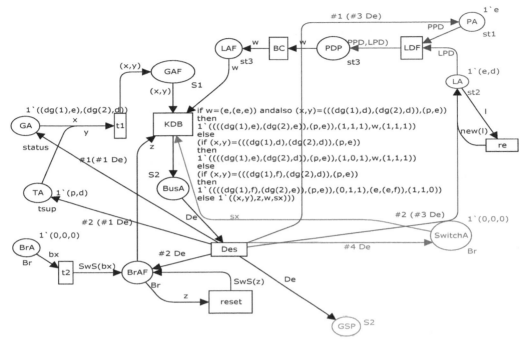

Figure 4. CPN of multi-agent power system

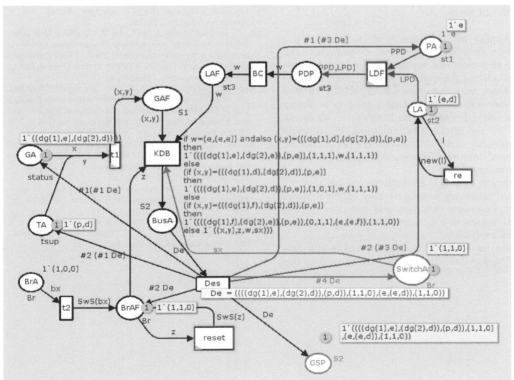

Figure 5. The tokens in places at major steps

5 SIMULATION RESULTS

The scenario simulation is provided as follows: the supply subsystem (SS) is consists of two diesels motors and gas turbine. The load system (LS) consists of the propeller and one load subsystem with two main consumers. Typical states are energized (e), de-energized (d) and fault (f). Circuit breakers and switches have two states: Open (1) and Close (0).

In this approach, the objective of inference is to solve load transfer problem in case of contingenous. The inference works as follows:

1 Initialize prime mover element statuses of PPS system as energized and the breakers/switches to be opened. Initialize the color settings of the CPN.
2 Evaluate all T node to form the enabled T-node set.
3 If the guard function of the enabled KDB node is evaluated to be true, the KDB node make decision and new supply plan.
4 Firing the activated KDB nodes by passing the tokens from the incoming P nodes to its outgoing P nodes -BusA.

At a certain time reference, the supply plans about all the generators and load are shown on Fig 5.

6 CONCLUSION

The purpose of simulation work is to develop a model on simulator for study the dynamic behavior of the ship electrical system, engine room or ship bridge. During the simulation we do some mistakes but we can learn from mistakes especially our own and share with others (during workshop or during the training). Training of students/ship crews can do reliability for certain situation using ship or ship bridge/engine room simulators. The lectures have developed training scenarios with particular attention to human factors.

Although Petri nets are more complicate than the standard reliability techniques they enable a richer insight in system behavior and becomes an important element of education for reliability engineers and risk analysts.

Since the whole approach is derived from a simplified ship system structure, the future work of this research will study more complex system structure. Agents control for synchronous generator, propulsion induction motor, and power inverter will be considered.

AKNOWLEDGMENT

The results presented in the paper have been derived from the scientific project "New technologies in Diagnosis and Control of Marine Propulsion Systems" supported by the Ministry of Science, Education and Sports of the Republic of Croatia."

REFERENCES

Dicesare, F.,. Desrochers, A. A.,1991. „ *Modeling Control and Performance Analysis of Automated Manufacturing Systems Using Petri Nets,"* Advances in Control and Dynamic Systems, C. T. Leondes (edition), vol. 45, Academic Press, pp. 121-172

Garcia, E., Rodríguez L., Moran F. Morant, Correcher A., Quile, E. , Blasco R.,2008. *Fault Diagnosis with Colored Petri Nets using* Latent Nestling Method, *Proc. of* ISIE08, Cambidge, UK

Jensen, K., L. M., Kristensen L. Wells,2007. *Colored Petri Nets and CPN Tools for Modelling and Validation of Concurrent Systems*, International Journal On Software Tools for Tech. Transfer (STTT), Springer- Verlag, vol. 9, Springer-Verlag, pp. 213-254.

Jun, Z., Junfeng L., Ngan H.W., Jie W., 2009. *A Multi-agent solution to energy management of distributed hybrid renewable energy generated system*, Proc. of 8th International Conference on Advances in Power System Control, Operation and Management, , HongKong

Krčum, M., Gudelj, A., Jurić , Z. 2005. *Shipboard Power Supply – Optimisation by Using Genetic   Algorithm.* Proceedings of the 11th IEEE International Conference on Methods and Models in Automation and Robotics, pp. 201 – 206, Miedzydroje, Poland

Kristensen, L.M. S., Christensen K. Jensen, 1998. *The Practioner's Guide to Coloured Petri Nets*, International Journal On Software Tools for Tech. Transfer (STTT), vol. 2, Springer-Verlag, 1998, pp. 98-132.

Murata. T., 1989. *"Petri nets: properties, analysis, and applications"*, in Proceedings of the IEEE, vol. 77, no. 4, pp. 541 – 580.

CPNTools, CPN Group, Department of Computer Science, University of Aarhus, Denmark. http://www.daimi.au.dk/CPnets/

19. Logical Network of Data Transmission Impulses in Journal-Bearing Design

K. Wierzcholski

Institute of Mechatronics, Nanotechnology & Vacuum Technique, Koszalin University of Technology, Poland

ABSTRACT: This paper presents the implementation of logical network of data transmission impulses in journal bearing optimum design regard to the journal-bearing operating parameters such as carrying load capacity, friction forces, friction coefficient and bearing wear.
Efficient functioning of slide journal-bearings systems occurring in maritime industry, require to choice the proper journal shapes, bearing materials, roughness of bearing surfaces and many other features to which belongs load carrying capacity and capability to the processes control. Artificial intelligence of micro-bearing leads to the creating and indicating of the logical network models of data transmission impulses, to describe most simple and most proper graphical schemes presenting the design of anticipated processes. Application of the logical network analysis into the slide journal bearing design is the subject-matter of this paper.

1 SYMBOLS

A,B,C,D,E,F-ingoing signals,
AI-Artificial Intelligence,
ICS-inverse Control System,
LNAS-Logical Network Analysis System,
LNA-Logical Network Analysis,
$\mathbf{U}$-input vector
Y-outgoing signals,
$\cup$-union mechanism,
$\cap$-intersection mechanism,
$\sim$ mechanism which negates each impulses

2 LOGICAL NETWORK OF IMPULSES IN SLIDE JOURNAL BEARING SYSTEM

Artificial intelligence supported by the logical network of data transmission impulses includes in a new technologies industry, and in many of the most difficult problems connected with optimum strategy of computer science program performances [1],[2]. Network analysis for Artificial Intelligence (AI) research is highly applied mathematical knowledge. Subfields of Logical Network Analysis (LNA) in the field of slide journal -bearing systems are organized around following particular problems:
- The creating of models of LNA for computer calculations taking into account the graphical scheme tools [6].
- The study of simplification of the Logical Network Analysis Scheme (LNAS) by virtue of

graphical topology and attempt to create of the equivalence of Logical Network Analysis Scheme (LNAS)$_{eq}$ using the logical set theory and topology laws [3],[4].
- The applications of logical network analysis to the intelligent control theory and cyber-bio-tribology.

3 CONTROL SYSTEM

Logical Network Analysis Scheme (LNAS) is a member of Artificial Intelligence (AI) of micro-bearing systems. LNAS is beginning from electronic input impulses and is finished in output electronic impulses.
Sometimes in particular cases we can observe the Inverse Control Systems (ICS) for example in structure of neural network [7] and learning algorithms occurring for example in bioreactor see Fig.1

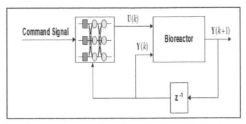

Figure 1. Control system based on ICS, where U- input vector, Y-output vector

In slide journal bearing systems we have the input vector **U**(A,B,C,D,…) in multi-dimension space. The components in such vector have various meaning. For example if we take into account the shape of the micro-bearing journal, then we have: A–cylindrical journal shape of the journal, B–conical journal shape of the journal, C–spherical journal shape of the journal, D–parabolic journal shape of the journal; E_n–various magnitudes of micro-bearing surface roughness for n=1,2,…; F_n– various magnitudes of radial clearances for n=1,2,…

In slide journal-bearing the output electronic impulse vector **Y** leads for example to such components as: load carrying capacity, friction force, friction coefficients, micro-bearing wear.

4 THE LOGICAL NETWORK TOOLS FOR IMPULSES SCHEME

Here are presented the tools of LNAS occurring in micro-bearing electronic network and in computer science [6],[7],[8].

We assume the following nods as connection boxes:

∪ – union (sum) mechanism,

∩ –intersection mechanism which choices common properties of two impulses,

~ –mechanism which negates each impulse.

Above mechanisms are presented and explained in Fig.2.

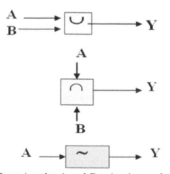

Figure 2. Input impulse A and B going into nods and output impulse Y outgoing from the nods

5 ABOUT THE TRANSMISSION OF IMPULSES FOR SLIDE JOURNAL BEARING DESIGN

5.1. Now for a one device we assume following expression:

Y(A,B,C,D)=

=[~(A∪C)∩B]∪{[(~C∩B) ∪D] ∩A},   (1)

where: A,B,C,D –input impulses, Y(A,B,C,D)-output impulse of first kind.

In practical cases we have a lot input impulses.

Tribo-topology scheme of $(LNAS)_1$ for the formula (1) is presented graphically in Fig.3.

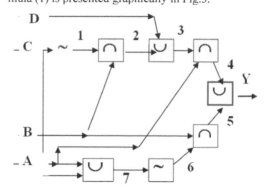

Figure 3. Tribo-topology logical network analysis scheme $(LNAS)_1$: Y=[~(A∪C)∩B]∪{[(~C∩B) ∪D] ∩A} for seven connections

By virtue of the set theory the expression (1) we can transform in following form [4]:

$$Y≡[~(A∪C)∩B]∪\{[(~C∩B) ∪D] ∩A\}$$
$$≡[~A∩(~C∩B)]∪\{[(~C∩B) ∩A]∪(A∩D) \} ≡$$
$$≡[~A∩(~C∩B)]∪[(~C∩B) ∩A]∪(A∩D) ≡$$
$$≡(~A∪A)∩(~C∩B)∪(A∩D) ≡$$
$$≡X∩(~C∩B)∪(A∩D) ≡$$
$$≡(~C∩B)∪(A∩D).   (2)$$

Hence we have finally:

$$Y(A,B,C,D)≡(~C∩B)∪(A∩D)   (3)$$

Symbol Y denotes total impulses space.

In Fig. 4 for A,B,C,D –input impulses, the output impulse Y treated as the result (3) is showed in the graphical form.

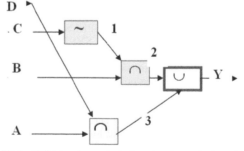

Fig.4. Tribo-topology logical network analysis scheme $(LNAS)_{1eq}$: Y≡(~C∩B)∪(A∩D) for three connections

It is visible that in this case the equivalent scheme $(LNAS)_{1eq}$ presented in Fig.4 is more simple than the origin scheme $(LNAS)_1$ presented in Fig.3 because $(LNAS)_1$ has seven connections between two variable nods (boxes), however $(LNAS)_{1eq}$ has only three. The numbers of connections denotes

146

more proper und more productive network of transmission impulses.

During the slide bearing and especially HDD micro bearing calculations we had been used and utilized the sequence of various mutually connected data of experimental AFM measurement results. The most simple geometry and sequence logical architecture of connections of used data impulses have the important influences on the convergence of computer calculations of necessary exploitation parameters. Hence follows the weightiness of the proposed solution in comparison to the others methods. More examples of the proposed system had been described in the last authors and his research team papers [8],[9],[10].

5.2.Now we are going to show the not effective case of the presented input electronic impulses method of second kind. We take into account the next output impulse:

$$Y(E_1, E_2, F_1, F_2),\qquad(4)$$

and we assume following expression:

$$Y=[\sim(E_1 \cap E_2) \cup F_1] \cap [\sim (F_1 \cup \sim F_2) \cap E_1],\qquad(5)$$

where: E_1, E_2, F_1, F_2 –input electronic impulses, Y-output impulse.

On the ground of the simple set theory we can transform expression (5) into the following form:

$$Y \equiv [\sim(E_1 \cap E_2) \cup F_1] \cap [\sim (F_1 \cup \sim F_2) \cap E_1] \equiv$$
$$\equiv (\sim E_1 \cup \sim E_2 \cup F_1) \cap (\sim F_1 \cap F_2 \cap E_1).\qquad(6)$$

Result is as follows:

$$Y(E1, E2, F1, F2) \equiv$$

$$\equiv (\sim E1 \cup \sim E2 \cup F1) \cap (\sim F1 \cap F2 \cap E1).\qquad(7)$$

A new Tribo-Topological scheme of network $(LNAS)_2$ described by the formula (5) is presented in Fig.5.

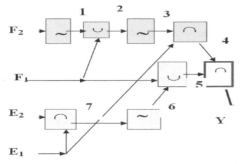

Figure 5.Tribo-topology logical network analysis scheme $(LNAS)_2$ for seven connections :
$Y=[\sim(E_1 \cap E_2) \cup F_1] \cap [\sim (F_1 \cup \sim F_2) \cap (E_1]$

In Fig. 6 is presented the equivalent scheme $(LNAS)_{2eq}$ described by the formula (7) for E_1, E_2, F_1, F_2 –input impulses, Y-output impulse.

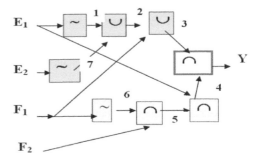

Figure 6.Tribo-topology logical network analysis scheme $(LNAS)_2$ for seven connections:
$Y \equiv (\sim E_1 \cup \sim E_2 \cup F_1) \cap (\sim F_1 \cap F_2 \cap E_1)$

It is visible that in this case the scheme $(LNAS)_1$ of origin network presented in Fig.5 and the equivalent scheme $(LNAS)_{2eq}$ presented in Fig 6 have the same number of connections. The arrow denotes the direction of impulse transmission.

6 NUMERICAL TOOLS

Fig.7 presents pressure distribution and loads carrying capacity in conical slide microbearing as a one component of the output vector by virtue of the numerical calculations [5] of modified Reynolds equation performed in Matlab 7.2 Professional Program.

$R = 0.001$ [m], $L_{c1}=b_c/R=1$, $\eta_o = 0.03$ [Pas], $\omega=565.5$ [1/s], $p_o=16.96$ [MPa],

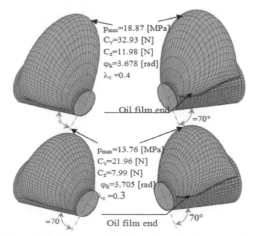

Figure 7. The pressure distributions in conical micro-bearings caused by the rotation in circumferential direction where conical inclination angle $\gamma=70°$.

Left side of figures 7 presents the view from the film origin, right side shows the view from film end. Output vector has following components: film end coordinates $\varphi_k=3.678$ for pressure in two upper distributions, φ_k=3.705 for pressure in two lower distributions, maximum pressure value p_{max}, load carrying capacity components C_z, C_y in z (longitudinal or axial) and y (transverse) directions. We have the following input impulses: relative eccentricity ratio λ_c=0.4 for two upper pressure distributions, relative eccentricity ratio λ_c=0.3 for two lower pressure distributions, radius of the cone R=0,001m, bearing length b_c, dimensionless bearing length L=1, angular velocity of the cone journal ω= 565.5 s^{-1} , oil dynamic viscosity η=0.03 Pas.

7 COROLLARIES

Corollary 1.*Final calculation indicator Y_{ID} is defined as a union of three elements namely of the two output impulses of two kinds of networks including various transmission impulses and the partial modified differential Reynolds equation describing pressure and load carrying capacity distribution. Such indicator is defined in the following form:*

$$Y_{ID}=Y(A,B,C,D) \cup Y(E_1,E_2,F_1,F_2)\cup Y. \qquad (8)$$
(partial differential equation).

Corollary 2.*The final calculation indicator with the sum of output impulses having the least sum of connections is a most productivity logical network of data transmission impulses in journal bearing design and in the case of identification of journal bearing work parameters using acoustic emission methods.*

Corollary 3. Taking into account influences of variable sometimes mutually depended impulses on the behavior of slide journal bearing system, we can investigate how the design variables of mentioned bearings affect the bearing stiffness and the natural frequencies of the bearing shaft [8],[9],10].

Corollary 4. This research also shows that the supporting structure which includes the stator, housing and base plate plays an important role in determining the natural frequencies and mode shapes of slide journal bearing system [8],[9],10].

8 CONCLUSIONS

Results obtained in this paper in the field of logical network analysis for data transmission impulses during journal bearing design presented in graphical form as a mathematical set theory implementation constitute a new convenient tools of artificial intelligence methods and computer calculation methods occurring in maritime transport [8],[9],10].

Presented paper establish the scheme of calculation algorithm of hydrodynamic pressure and carrying capacity changes in journal-bearings for various journal shapes and for various geometries.

The above mentioned results enable to investigate the dynamic behavior of journal bearing system by solving Reynolds equation and the equations of motion in some degree of freedom. It shows that the dynamic journal bearing can be affected not only by the design variables but also by already existing motor parameters [8],[9],10].

ACKNOWLEDGEMENT

This paper was supported by Polish Ministerial Grant 3475/B/T02/2009/36 in years 2009-2012.

REFERENCES

[1] Bharat Bhushan: Nano-tribology and nanomechanics of MEMS/NEMS and BioMEMS, BioNEMS materials and devices. Microelectronic Engineering 84, 2007, pp.387-412

[2] Jang G. H., Seo C.H., Ho Scong Lee: Finite element model analysis of an HDD considering the flexibility of spinning disc-spindle, head-suspension-actuator and supporting structure. Microsystem Technologies, 13, 2007, pp.837-847

[3] Kącki E.:Partial differentia equationsin physical and technical problems (In polish).WNT Warszawa 1968

[4] Kuratowski K.: Introduction into set thepry and topology (in polish) PWN Warszawa 1970,

[5] Ralston A:A First Course in Numerical Analysis (in Polish),PWN,Warszawa 1971

[6] Wierzcholski K.: Enhancement of memory capacity in HDD micro- bearing with hyperbolic journals, Journal of Kones Powertrain and Transport, Warsaw 2008, Vol.15, No.3, pp.555-560

[7] Wierzcholski K.: Bio and Slide Bearings, their Lubrication by Non-Newtonian Oils and Applications in Non-Conventional Systems, Vol.1, Gdansk Univ.of Technology, GRANT UNI EU: MTKD-CT-2004-517226

[8] Wierzcholski K.: Fuzzy Logic Tools in Intelligent Micro-Bearing Systems.(in English). XIII Journal of Applied Computer Science, vol.17 No.2,2009, pp.123-131

[9] Wierzcholski K., Chizhik S., Trushko A., Zbytkowa M., Miszczak A.: Properties of cartilage on macro and nano-level. Advances in Tribology, vol. 2010, Hindawi Publishing Corporation, New York: http://www.hindawi.com/apc.aspx?n=243150

[10] Wierzcholski K., Chizhik S., Khudoley A., Kuznetsova K., Miszczak A.: Micro and Nanoscale Wear Studies of HDD Slide Bearings by Atomic Force Microscopy. Proceedings of Methodological Aspects of Scanning Probe Microscopy,Heat and Mass Transfer Institute of NAS, Minsk 2010, pp. 247-252

Ship Propulsion and Fuel Efficiency
Miscellaneous Problems in Maritime Navigation, Transport and Shipping – Marine Navigation and Safety of Sea Transportation – Weintrit & Neumann (ed.)

20. Optimum Operation of Coastal Merchant Ships with Consideration of Arrival Delay Risk and Fuel Efficiency

K. Takashima
Tokai University, Shizuoka, Japan

B. Mezaoui
Graduate School, Tokyo University of Marine Science and Technology, Tokyo, Japan

R. Shoji
Tokyo University of Marine Science and Technology, Tokyo, Japan

ABSTRACT: In this study the authors proposed a weather routeing algorithm that takes into account the risk of delay on arrival time due to the incertitude in the wind and wave forecasts for the computation of an optimal minimum fuel route allowing the ship to reach the destination port at the scheduled arrival time.
By investigating the accuracy of wind and wave forecasted data during a determined period, the characteristics of variation of forecasts errors and the correlations between the various forecast errors were determined. Using the hence acquired knowledge a method for estimating the standard deviation of the arrival time error and the risk of delay for a minimum fuel route using Dijkstra's algorithm was proposed. The effectiveness of the proposed method was verified by carrying out simulations.

1 INTRODUCTION

The efficiency of weather routeing (here after WR) for the safety and fuel consumption reduction has been largely documented and demonstrated. The reliability of the Optimal Route computed using WR depends largely on the accuracy of the weather and ocean conditions and surface oceanic current data. With the recent development of the weather forecast technique this reliability increased and the results of fuel consumption reduction during voyage has been increasing.

For the coastal shipping fuel consumption saving is also important, however more than that is the importance of reaching the destination port at the scheduled time without any delay. In general coastal merchant vessels have limited power engines, due to that masters of such ships sail with full engine to arrive at their destination port as soon as possible to prevent the possibility of losing time if encountering any possible bad weather during the voyage, in which case with their limited engine power they will not be able to withstand the bad weather conditions and their arrival would be delayed. By closing to their destination port the masters reduce their engine to adjust their arrival time.

The authors think that by analysing the weather and ocean conditions over a given and long enough period of time and determining the forecast errors and their evolution in space and time, we would be able to use this acquired knowledge to estimate the error on arrival time. Furthermore we would be able to estimate the appropriate cruising speed that would

allow us to reach the destination port with a predetermined arrival time error percentage, avoiding running the engine at higher rpm than necessary and hence saving fuel. Such information would be very useful for coastal shipping.

In this research, a method for estimating the standard deviation on the arrival error for the coastal merchant vessels calculated using a minimum fuel routing algorithm is proposed. The accuracy of the wind and wave forecast data used is also investigated. And then simulation for coastal sailing are carried out to show the efficiency of the proposed method.

2 MFR WITH CONSIDERATION OF ARRIVAL DELAY RISK

In this part the MFR method applied for the coastal vessel is explained, the wind and wave forecast accuracy investigated and the engine performance and response to the weather conditions are shown. The algorithm for reaching the destination with a predetermined percentage is also explained (here after this route will be noted to as MFR considered risk of delay).

2.1 *Method of calculation of MFR with consideration of risk of delay*

In this research the algorithm adopted for the MFR routeing of the coastal vessel is one the greedy algorithm methods, known as Djikstra's algorithm. The

MFR with consideration of risk of delay is computed according to the following method.

1 First compute the MFR_0 to reach the destination port at the scheduled time T.
2 For the computed MFR_0 estimate the standard deviation of the arrival error σ_T (the probability for reaching the destination port at after the scheduled time T with a probability of σ_T, $2\sigma_T$, $3\sigma_T$ is respectively).
3 Since the wind and wave forecast are uncertain, we introduce the concept of risk of delay to our problem, we recompute the MFR route to reach the destination point before the time T with a determined time $k\sigma_T$ where k is the weight given to the risk delay and lets call this newly recomputed route as MFR'_0, $k\sigma_T$ is called the time margin.
4 Run the ship on this newly recomputed MFR'_0 until the wind and wave forecast are updated, at that time compute from the reached position the MFR_1 to reach the destination port, and keep repeating (2), (3), (4) until reaching the destination point.

With the ship approaching its destination port the wind and wave forecast period becomes shorter and shorter and the standard deviation σ_T becomes smaller, hence the delay risk time $k\sigma_T$ is sufficiently reduced and the destination port can be reached at the scheduled time.

2.2 Estimation method of the standard deviation of arrival time error

Here the method of estimation of the standard deviation σ_T on the arrival time of the computed MFR is presented. Due to the complexity of estimating and analysing at the same time the forecast accuracy for the wind, wave and surface oceanic currents data necessary for estimating the σ_T, only the accuracy information for the wind and wave forecast data was taken into consideration.

Consider a ship leaving the departure point P_0 at time T_0 and sailing toward the destination point P_f (Figure 1). The ship is assumed to sail with a constant heading for an interval of time of Δt. The ship will reach the $i+1^{th}$ grid waypoint P_{i+1} at the time T_{i+1} which can be written as:

$$T_{i+1} = T_i + \frac{D_i}{V_{gi}} \quad (i = 0, 1, 2, ---) \tag{1}$$

Here V_{gi} is the speed over ground of the ship at waypoint P_i, D_i is the distance separating the two waypoints P_i and P_{i+1}. The speed over the ground of the ship is function of the ship's speed through the water, the drift angle due to the blowing wind and the ocean current vector, so we can write:

$$V_{gi} = V_{gi}(V_i, \alpha_i, W_i) \tag{2}$$

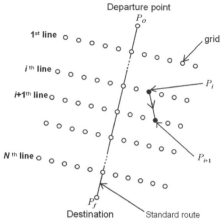

Figure 1. Grid points of Dijkstra's algorithm to calculate minimum fuel route.

With V_i representing the speed through the water, α_i is the drift angle and W_i the ocean current vector.

The ship's speed through the water can be written as:

$$V_i = V_i(\theta_i, n_i, S_i) \tag{3}$$

where V_i is the speed through the water θ_i is the ship's heading and n_i is the propeller revolution number between waypoint P_i and waypoint P_{i+1}.

The vector S_i representing the wind and wave conditions is noted as:

$$S_i = \begin{bmatrix} w_i & \gamma_i & h_i & \varphi_i & F_i \end{bmatrix}^T \tag{4}$$

where $w_i, \gamma_i, h_i, \varphi_i$ and F_i denote the wind speed, wind direction, significant wave height, predominant wave direction and average wave period at the ship's position P_i at the time T_i respectively.

ΔT_{i+1} the error on forecasted time arrival at time T_{i+1} can be written as:

$$\Delta T_{i+1} = \Delta T_i + \frac{-D_i}{V_{gi}^2} \Delta V_{gi} \tag{5}$$

By assuming that the error on the drift angle is null and that the ocean current contains no error, ΔV_{gi} the error on the speed over ground V_{gi} at time T_i can be written as:

$$\Delta V_{gi} = \frac{\partial V_{gi}}{\partial V_i} \Delta V_i = \cos\alpha_i \Delta V_i \tag{6}$$

By replacing equation (6) in (5) we get:

$$\Delta T_{i+1} = \Delta T_i + \frac{-D_i}{V_{gi}^2} \cos\alpha_i \Delta V_i \tag{7}$$

By rewriting equation (3) and expanding it we can write:

$$\Delta V_i = \frac{\partial V_i}{\partial S_i} \Delta S_i$$

$$= \begin{bmatrix} \dfrac{\partial V_i}{\partial w_i} & \dfrac{\partial V_i}{\partial \gamma_i} & \dfrac{\partial V_i}{\partial h_i} & \dfrac{\partial V_i}{\partial \varphi_i} & \dfrac{\partial V_i}{\partial F_i} \end{bmatrix} \begin{bmatrix} \Delta w_i \\ \Delta \gamma_i \\ \Delta h_i \\ \Delta \varphi_i \\ \Delta F_i \end{bmatrix} \tag{8}$$

Replacing (8) in (7) gives:

$$\Delta T_{i+1} = \Delta T_i + \frac{-D_i}{V_{gi}^2} \cos\alpha_i \frac{\partial V_i}{\partial S_i} \Delta S_i \tag{9}$$

ΔS_i the forecast error on wind and wave vector is not simply a Gaussian purely random sequence (Gaussian white sequence with a zero mean); in general, there exists a relation between the forecast errors ΔS_{i+1} and ΔS_i. Thus we can regard ΔS_i as Gauss-Markov random sequence produced by the following shaping filter:

$$\Delta S_{i+1} = \Phi_i \Delta S_i + \Delta S_{i+1}^* \tag{10}$$

where ΔS_{i+1}^* is a Gaussian white sequence with zero mean value, Φ_i is the normalized correlation matrix representing the degree of correlation between the two forecast error vectors ΔS_i and ΔS_{i+1} at times T_i and T_{i+1} respectively .

Combining equation (9) and (10) we obtain the following vectorial equation:

$$\begin{bmatrix} \Delta T_{i+1} \\ \Delta S_{i+1} \end{bmatrix} = \begin{bmatrix} 1 & -\dfrac{D_i}{V_{gi}^2} \cos\alpha_i \dfrac{\partial V_i}{\partial S_i} \\ 0 & \Phi_i \end{bmatrix} \begin{bmatrix} \Delta T_i \\ \Delta S_i \end{bmatrix} + \begin{bmatrix} 0^T \\ I \end{bmatrix} \Delta S_{i+1}^* \tag{11}$$

where, I is the identity matrix.

Equation (11) can be written in a more simplified form as:

$$\Delta Z_{i+1} = A_i \Delta Z_i + B \Delta S_{i+1}^* \tag{12}$$

The covariance matrix M_{i+1} of the error vector ΔZ_{i+1} is given by:

$$M_{i+1} = A_i M_i A_i^T + B Q_{i+1} B^T \tag{13}$$

where, Q_{i+1} is the covariance matrix of ΔS_{i+1}^*.

Substituting on both sides of equation (10) and taking the expected values we get:

$$P_{S_{i+1}} = \Phi_i P_{S_i} \Phi_i^T + Q_{i+1} \tag{14}$$

where, $P_{S_{i+1}}$ is the covariance matrix for ΔS_{i+1} and P_{S_i} the covariance matrix for ΔS_i and Φ_i is the correlation matrix.

Q_{i+1} the covariance matrix of ΔS_{i+1}^* can be written as:

$$Q_{i+1} = P_{S_{i+1}} - \Phi_i P_{S_i} \Phi_i^T \tag{15}$$

From that the covariance matrix M_{i+1} of the error vector ΔZ_{i+1} can be written as:

$$M_{i+1} = A_i M_i A_i^T + B(P_{S_{i+1}} - \Phi_i P_{S_i} \Phi_i^T) B^T \tag{16}$$

The upper left term of the covariance matrix M_{i+1} is the standard deviation $\sigma_{T_{i+1}}^2$ of the arrival time error at waypoint P_{i+1}.

Using equation (16) we can estimate the standard deviation of the arrival error at each waypoint at each time ($i = 0, 1, 2, -----$) until the arrival waypoint is reached, and so the standard deviation σ_T on the arrival time of the computed MFR can be estimated.

3 ACCURACY OF WIND AND WAVE FORECASTS FOR THE NAVIGATION AREA

For computing the MFR with risk of arrival delay, Dijkstra's algorithm was used; each time rerouting was done using the above mentioned method the standard deviation on the arrival time predicted error was computed. For computing this standard deviation information about the accuracy of wind and wave forecasts is necessary. Since the main factors affecting the ship's engine performance and fuel consumption are the wind and wave conditions, information about the degree of their forecasts accuracy is really important.

In this research the accuracy of wind and wave forecasts produced each 3 hours basis time (00, 03, 06, 09, 12, 15, 18, 21UTC) with a spatial resolution of the forecasts is 2′ latitude by 2′ longitude,forecatsed for every hour up to a period of forecast of 72 hours ahead were investigated for a period of one month, the. The mean average error and the covariance matrix for each parameter (wind speed, wind direction, significant wave height, predominant wave direction and average wave period) were computed for every grid point for every hour. The correlation coefficients between the different parameters forecasts errors were also calculated. The analysis issued each 3 hours basis time was considered to be representing the actual state of each parameter and containing no analysis error.

3.1 *Mean and standard deviation of the wind and wave forecast errors*

For each grid, the mean errors (forecast bias) $E[\Delta \hat{C}(m,n,t)]$ of forecasted wind and wave data were calculated for every hour, up to 72 forecast hours according to the following equation:

$$E[\Delta \hat{C}(m,n,t)] = E[C(m,n) - \hat{C}(m,n,t)] \tag{17}$$

where:

m, n: the grid point number
t: the forecast time ($t=3,---,72$)

$C(m, n)$: the actual wind and wave conditions at grid (m, n)

$\hat{C}(m,n,t)$: the forecasted wind ad wave conditions at grid (m, n) at time t

E: the expected value (in this research it represents the average over the one month of the study)

The components of vector $C(m, n)$ and $\hat{C}(m,n,t)$ are as follows:

$$C(m,n) = \left[w_{wi} \gamma_{wi} h_{wv} \varphi_{wv} F_{wv} \right]^T \qquad (18)$$

$$\hat{C}(m,n,t) = \left[\hat{w}_{wi} \hat{\gamma}_{wi} \hat{h}_{wv} \hat{\varphi}_{wv} \hat{F}_{wv} \right]^T \qquad (19)$$

Where w_{wi}, γ_{wi}, h_{wv}, φ_{wv}, F_{wv} are the actual wind speed, wind direction, significant wave height, predominant wave direction and average wave period respectively at grid(m, n) and $\hat{w}_{wi}$, $\hat{\gamma}_{wi}$, $\hat{h}_{wv}$, $\hat{\varphi}_{wv}$, $\hat{F}_{wv}$ are the forecasted wind speed, wind direction, significant wave height, predominant wave direction and average wave period respectively at grid(m, n) at forecast time t.

For obtaining the unbiased wave ad wind forecasted data $\overline{C}(m,n,t)$ the mean error of the forecasted wind and wave data $E[\Delta \hat{C}(m, n, t)]$ is added to forecasted wind and wave data $\hat{C}(m,n,t)$.

$$\overline{C}(m,n,t) = \hat{C}(m,n,t) + E[\Delta \hat{C}(m,n,t)] \qquad (20)$$

The errors of the unbiased wind and wave forecasted at each grid point are defied as:

$$\Delta C(m,n,t) = C(m,n) - \overline{C}(m,n,t) \qquad (21)$$

Multiplying the error vector $\Delta C(m,n,t)$ by its transpose and averaging it over the study period the covariance matrix of the errors in the unbiased wind and wave forecasted data $P_C(m,n,t)$ can be computed by:

$$P_C(m, n, t) = E[\Delta C(m, n, t)\Delta C(m, n, t)^T] \qquad (23)$$

The diagonal terms of the covariance matrix P_c represents the variance errors in the forecasted wind speed, wind direction, significant wave height, predominant wave direction and average wave period respectively.

To investigate the difference in accuracy of wind and wave forecasted data over the simulation sea area, the area of simulation was divided into sub-area of $30' \times 30'$ as shown in figure 2. If there is no land in any sub-area, it should contain 255 grid points. The mean forecast error and the average of the standard deviation of the forecast error were calculated for each sub area.

Figure 3 shows the mean and standard deviation of the forecast error for the significant wave height for 5 different sub-area previously shown in figure 2.

The mean errors are shown in continuous lines and the standard deviations in dashed lines. The mean forecast error varies in the range of ±0.2m for all the selected sub-area, and has a negative value for the period of forecast extending up to 45-55 hours ahead. As the forecast time becomes larger the mean error changes of sign and becomes positive, which imply that the significant wave height is over-estimated for the first 2 days of the forecast period and underestimated for above that period. The standard deviation increases as function of forecast time.

The same checking was also performed for the other parameters, namely wind speed, wind direction, predominant wave direction and average wave period, results show the same pattern for all the different parameters with a standard deviation increasing as function of forecast time, meanwhile no significant differences were found between the different investigated sub-area.

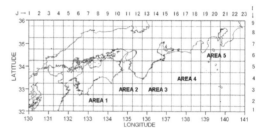

Figure 2. Sub-area of wind/wave forecasts.

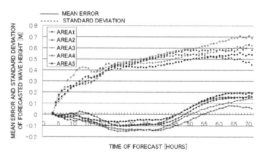

Figure 3. Mean error and standard deviation of forecasted wave height averaged over each sub-area versus the time of forecast.

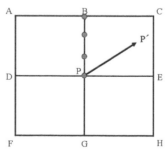

Figure 4. Image chart to calculate correlation.

152

3.2 Correlation between the wind and wave forecast errors

For a ship making way, there is correlation between the forecast error of wind and wave observed at given position and time and that observed at position after sailing for a Δt time. For investigating that; the correlations coefficients between the forecast errors at time t for every grid point and its surrounding grid points at time (t+Δt) were calculated.

As shown in Figure 4, for a given grid point P, the correlations coefficients between the forecast errors at point P and time t and the forecast errors at grid point A at time (t+Δt) were calculated. The same has been done for the all the other surrounding grid points from B to H. For the purpose of the simulation, in which the position of the ship is computed every 15 minutes, Δt was set to 15minutes; the distance between the central grid point and the surrounding grid points was set to be equal to the maximum distance that the ship can sail during the time Δt.

Figure 5 shows the correlations coefficients between the forecasts errors for wind speed averaged over the sub-area 3 as function of forecast time. The directions 000°, 090°, 180°, 270° denotes the direction of the second grid point from the first grid point. Meridional correlation coefficients are shown in continuous line, whereas the zonal correlation coefficients are shown in dotted lines. It is clear from figure 5 that the zonal correlation coefficients are larger than the meridional ones. The same pattern also applies for all the sub-area from 1 to 5.

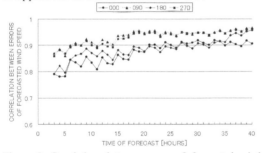

Figure 5. Correlations between errors of forecasted wind speeds as function of forecast time (for sub-area 3).

4 SIMULATION FOR THE MFR WITH CONSIDERATION OF RISK OF DELAY

Both the results obtained from the analysis of the wind and wave forecasts errors and the method for computing the standard deviation of the arrival time error were used to carry out MFR simulation with consideration of risk of delay. In the simulation the wind and wave forecasted data issued just before the departure were used to compute a MFR and the standard deviation σ_T of the arrival time error for that MFR. After considering the risk of delay a new MFR reaching the destination $2\sigma_T$ ahead of the scheduled time is recomputed. This $2\sigma_T$ time is called the allowance time. As the forecast data is updated the MFR and the allowance time are recomputed.

4.1 Mean and standard deviation of the wind and wave forecast errors

For the purpose of simulation a real ship, a cement carrier plying on the route Ube-Nagoya/ Tokyo was used as a model ship. Table1 shows the main characteristics of the model ship.

Table 1. Principal particulars of the model ship.

Length over all	159.7 m
Length between perpendiculars	152.5 m
Breadth	24.2 m
Full load draught	9.016 m
Gross tonnage	13,787 ton
Engine type	Diesel engine x 1
Max engine power	6,960 kW
Sea speed	13.0 kn

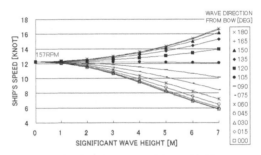

Figure 6. Speed performance curves of the model ship.

The speed performance curve was obtained from water tank trial results and real data obtained during sailing. The effect of wind and wave forecast errors on the ship's speed through the water error is shown in equation (8).

The hourly fuel consumption of the model ship is given by:

$$F=K\,P \qquad (24)$$

where:

K: the specific fuel consumption, for our model ship K=0.182kg/(kW·hour)

P: the engine output in KW

Figure 6 shows the ship's speed performance curve drawn for 157 rpm and for a wave direction from the bow of 15°.

4.2 *Verification of the effectiveness of the MFR with consideration of risk of delay*

To check the results of the simulations carried out, the results obtained when sailing the ship on its usual route were compared with the results obtained when sailing the ship on the MFR obtained by the proposed method.

Simulation is carried from the departure point of Bungo Suido until the arrival point at Uraga Suido, with the following conditions.

– Departure time: 21st Nov. 2006 09:06(UTC)
– Departure position: 33°01.73′N, 132°08.50′E
– Arrival time: 22nd Nov. 2006 20:21(UTC)
– Arrival postion : 35°06.52′N, 139°44.89′E
– Voyage time: 35.26 hours

Figure 7 shows the MFR computed at the departure time and the ocean currents chart. From this figure we can notice that on the opposite of the actual (usual) route that passes close to shore, the computed MFR sails the ship through area with favourable current. Fuel consumption when sailing along the usual route is 20.14 tons, whereas the fuel consumption for the computed MFR is 18.45 tons, which represents a 8.38% fuel consumption saving.

The wind and wave forecast data are updated every 3 hours, so a new MFR needs to be recomputed every 3 hours, this process of recalculation is called rerouting. Figure 8 shows all the MFR with consideration of risk of delay obtained for all the rerouting during the voyage. The ○ marks show the position of the rerouting points, we can notice that there is not a large difference between all the computed MFRs, we can say that rerouting does not affect largely the MFR.

Figure 9 shows the room time (bar graph) variation and the propeller revolution number as function of recalculation number, the propeller revolution number for MFR with consideration of risk of delay is shown in continuous line with ● mark, whereas the dotted line with Δ shows the RPM for the MFR without consideration of risk of delay.

If the risk of delay is taken into consideration we get at the beginning and room time of 0.7 hours (2% of the total voyage time) and the necessary RPM is 156.16 rpm which is slightly 2 rpm higher than the 154.35 rpm obtained for the MFR without consideration of risk of delay. However by closing to the arrival point the allowance time becomes smaller and smaller and RPM is gradually reduced. On the other hand for the MFR not taking into consideration the risk of delay, the RPM is gradually increased during the voyage and the ship is not able to reach the destination at the scheduled time. The MFR algorithm taking into consideration the risk of delay is able to estimate the allowance time during the voyage and is able to reduce the risk of delay at each rerouting.

Regarding the fuel consumption; the fuel consumption for the MFR with consideration of risk of delay is 18.37 tons and is just slightly smaller than that for the MFR without consideration of risk of delay 18.37 tons. However this fuel consumption is still 8.79% smaller than that for the actual route, which proves the effectiveness of both the MFR with and without the consideration of risk of delay.

Three other voyages were simulated for the period of October 2006 during which the wind ad wave forecast accuracy was investigated. Results show that the MFR with consideration of risk of delay always achieve a fuel consumption saving, with a maximum of 18.7%. The largest allowance time estimated at the beginning of the computation was 0.87 hour, which represents 2.6% of the total voyage time, and through rerouting the allowance time was gradually reduced allowing to reach the destination at the scheduled time.

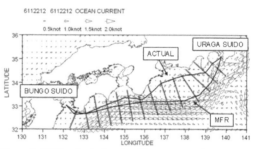

Figure 7. MFR and actual route with ocean current.

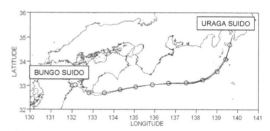

Figure 8. MFR considering or arrival delay risk.

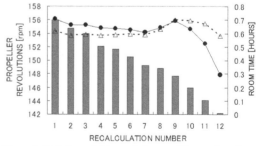

Figure 9. Risk time of delay and change of the number of propeller revolutions during the voyage.

5 CONCLUTION AND FUTURE WORK

In this research, by estimating the error on the estimated arrival time and its standard deviation, the efficiency of a MFR with consideration of risk of delay for the low fuel consumption operation of coastal merchant vessel has been demonstrated. First, based on the mean forecast errors for wind and wave forecasts, their distribution and variation, as well as the correlation between the different forecast errors a method for computing the standard deviation of the error on arrival time has been proposed. Considering the standard deviation on arrival time as risk of delay, it is possible to operate the ship without fear of arrival delay and at the same time reducing fuel consumption. Simulations were carried out using a model ship to verify and demonstrate the reliability and efficiency of the proposed method.

For the route from Ube to Tokyo, simulations were carried for 4 voyages, for all the four voyages, the allowance time estimated at the beginning of the computation gradually reduced during the voyage and the ship was able to reach the destination slightly ahead of the scheduled arrival time, and at the same time the fuel consumption was significantly reduced.

Taking in consideration the in exactitude of the wind and wave forecasted data, setting a given allowance time and reducing this allowance time throughout the voyage by reducing the propeller revolution number in such a way as to reach the destination at the scheduled arrival time is a feasible and practicable solution for weather routeing. Simulations of MFR with consideration of risk of delay demonstrated the feasibility of such a method.

For future development of this study and for the real application of the proposed method for the energy saving operation of coastal merchant vessel with consideration of the risk of delay experimentation with a real ship is scheduled. For the preparation to this experiment wind and wave forecast data have to be collected and analyzed over a long period to improve the information on the accuracy of the forecasted data for the various seasons and regions.

REFERENCE

Hagiwara. H & et. al. 2001. A Study on Stochastic Weather Routing –Accuracy of Estimated Standard Deviations of Passage Time and Fuel Consumption-. *The Journal of Japan Institute of Navigation vol.* 104: 1-11(in Japanese)

Hagiwara. H. 1982. A Method of Estimating the Error in Time Enroute Considering the Accuracy of Wave Prediction. *The Journal of Japan Institute of Navigation vol.* 67: 137-147(in Japanese)

Hagiwara. H. 1990. Stochastic Weather Routing Based of the Information of the Accuracy of Predicted Ocean Environments. *The Journal of Japan Institute of Navigation vol.* 83: 155-167(in Japanese)

Takashima, K. & et al. 2009. On the fuel saving operation for coastal merchant ships using weather routing. *8th International symposium on marine navigation and safety of sea transportation*: 431-436

Kano. K. & et al. 2008. Energy Saving Navigation Support System for Coastal Vessels. *Papers of National Maritime Research Institute. vol.*4. 35-72(in Japanese)

Dijkstra, E. W. 1959. A Note on Two Problems in Connexion with Graphs. *Numerische Mathematik 1*: 269-271

21. Digital Multichannel Electro-Hydraulic Execution Improves the Ship's Steering Operation and the Safety at Sea (Security of the Navigation Act)

Şt. Dordea
Constanta Maritime University, Constanta, Romania

ABSTRACT: Many electro-hydraulic drives use proportional valves or PWM (switching) valves, in order to obtain a linear response of the electric output or a good transfer of energy. The here-by presented solution combines the advantages of the two above-mentioned valves within a compartment of an electro hydraulic executor, being (cap)able to be digitally activated with a very good transfer of energy. The Multichannel Electro Hydraulic Executor (MEHE[1]) is a discrete/proportional device. The Digital Multichannel Electro Hydraulic Execution allows the data transfer within the entire electric steering control subsystem providing a safely ship's steering and increasing the navigation security. The multiway admission makes possible the rudder machine operation even if one or more channels become defective. Using individual pumps on each way the steering safety is increasing.

1 PREAMBLE

1.1 *The multichannel electro-hydraulic executor*

Many electro-hydraulic drives use proportional valves or PWM (switching) valves, in order to obtain a linear response of the electric input or a good transfer of energy.

The here-by presented solution combines the advantages of the two above-mentioned achievements within a compartment of an electro hydraulic activator, being (cap)able to be digitally activated with a very good transfer of energy.

Starting of the Linear Hydraulic Activator (cylinder) construction, one can imagine that the oil admission inside a compartment have place by many channels ($Ch_1 \ldots Ch_m$), under a constant pressure on each of these channels. The admission control is made by means of n elementary electromagnetic valves ($E_1 \ldots E_n$), as presented in Figure 1.

The elementary flow on each active way corresponding to a channel (it must be considered that the admission areas are identical - S_e) is Q_e. having a single active way, the flow dept will be:

$$Q_1 = Q_e = S_{pist} \cdot v_e,$$

where S_{pist} is the piston (spool) area;

v_e is the elementary speed of the spool.

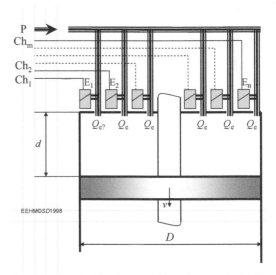

Figure 1. The principle of the multichannel electro hydraulic executor

Activating two ways, the flow will be double as before:

$$Q_2 = 2Q_e = 2S_{pist} \cdot v_e = v_2 \cdot S_{pist},$$

doubling the spool speed ($v_2 = 2v_1$).

Activating m channels, where $m \le n$, the flow will be:

---

[1] MEHE is a concept developed since 1998 by Dr Stefan Dordea (see the 2002 doctoral thesis "*The energy transfer within naval electro-hydraulic steering systems*")

$$Q_m = \sum_{m=1}^{n} S_{pist} \cdot v_e = S_{pist} \sum_{1}^{m} v_e = S_{pist} m v_e$$

The corresponding speed of the spool will be:

$$v_m = m \cdot v_e$$

The maximum speed of the spool will be $v_n = n \cdot v_e$, where n is the total number of ways.

Let's consider now an executor having 7 channels, each of them providing in active state the same elementary flow. The spool speed depending on the number of active ways is shown in Figure 2.

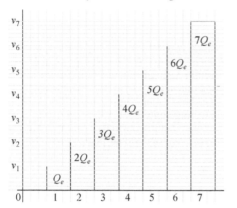

Figure 2. Linear speed of spool depending on the number of active ways

Special application needs no proportional dependence between number of ways and spool speed. So is the case presented in Figure 3.

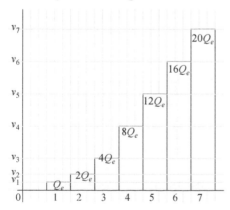

Figure 3. Non-linear spool speed depending on the number of active ways

1.2 Bi-Directional Multichannel Electro-Hydraulic Executor

Starting of the Multichannel Electro-hydraulic Executor construction, one can imagine a *bi-directional* device (Figure 4).

In order to ensure the motion of the piston (spool) towards both directions, two *electro-hydraulic multichannel ports* (Ch$_{S1}$...Ch$_{Sn}$ and Ch$_{P1}$...Ch$_{Pn}$, respectively) must be provided. Letter 'S' is associated to one motion direction and letter 'P' is associated to the other.

Activating the electromagnetic valves E1, respectively E2 provide the oil return, by means of the OR-P, respectively OR-S gates.

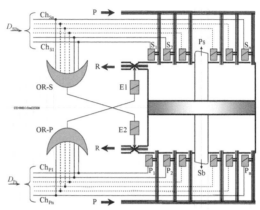

Figure 4. Linear speed of spool depending on the number of active ways

The electro-hydraulic device presented in Figure 4 together with the two OR gates must be seen as an entity conceived by the author of this paper work.

Any of the 'S' channels (Ch$_{S1}$...Ch$_{Sn}$) should be activated, the OR-S gate drives the electromagnetic valve E2. Any of the 'P' channels (Ch$_{P1}$...Ch$_{Pn}$) should be activated, the OR-P gate drives the electromagnetic valve E1.

The spool (piston) speed is depending on the number of activated channels:

$$v_m = \frac{Q_m}{S_{pist}} = \frac{\sum_{m} Q_e}{S_{pist}} = \frac{m Q_e}{S_{pist}}$$

The hydraulic energy power source (pump P) provides a constant and sufficient pressure on each channel, nondependent upon the asked flow by the device. For simplicity, there are no pressure regulators on each way marked on the drawing.

Designing the OR gates should take into consideration the number of channels used by the device. The returning flow provided by the electromagnetic valves E1, E2 must be at list equal to the sum of the maximum number of channels flow:

$$Q_{admE1,E2} \geq \sum_{e=1}^{n} Q_e = n \cdot Q_e$$

The drive accuracy is depending on the number of the channels. A higher number of channels makes the executor response almost linear, even the device is a discreet one. A higher number of channels means in the same time the increasing number of the elementary electromagnetic valves of the device, increasing the cost.

Figure 5 shows the dependence of the spool speed (for durations higher than the transitory ones) upon the total flow due to selective activation of the elementary electromagnetic valves, on one motion direction.

Subscript S of the speed symbol is assigned to Starboard side, Subscript P of the speed symbol is assigned to Port side

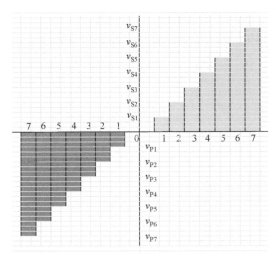

Figure 5. Speed versus debt depending on the number active channels

On the abscise axis one must consider the Q_{n-1} value between Q_{n-1} and Q_n.

Figure 3 presents the response of a bi-directional multichannel electro-hydraulic executor containing 2x7 channels of activation and identical areas of the orifices of the elementary electromagnetic valves of the device.

The flow corresponding to a single activated channel is:

$$Q_1 = S_{pist} \cdot v_e$$

The flow corresponding to $n-1$ activated channels is:

$$Q_{n-1} = \sum_1^{n-1} S_{pist} \cdot v_e = S_{pist} \sum_1^{n-1} v_e = S_{pist}(n-1)v_e$$

The maximum flow will be:

$$Q_n = \sum_1^n S_{pist} \cdot v_e = S_{pist} \sum_1^n v_e = S_{pist} \cdot nv_e$$

2 DIGITAL MULTICHANNEL ELECTRO-HYDRAULIC STEERING SYSTEM

2.1 The architecture of the digital Multichannel Electro-Hydraulic Steering System

Two links between the operational center and the rudder room are installed either by electric cables or by optic fibers (Figure 6).

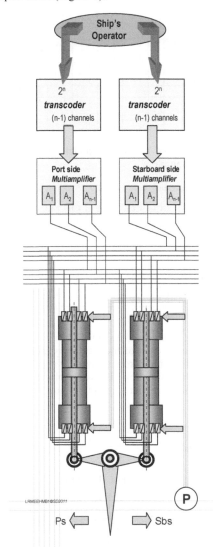

Figure 6. Multichannel activated Naval Steering System

There are digital signals to be converted into binary levels acting onto the corresponding channels.

2.2 *Transcoders*

As their name are suggesting, operational blocks making possible the conversion 2^n into $n-1$ channels are the transcoders.

Table 1. Transcoding truth table for 2n address to (n-1) channels

D				Ch 1	Ch 2	Ch 3	Ch 4		Ch n-2	Ch n-1
1	0	0	0	0	0	0	0		0	0
2	0	0	1	1	0	0	0		0	0
3	0	1	0	1	1	0	0		0	0
4	0	1	1	1	1	1	0		0	0
5	1	0	0	1	1	1	1		0	0
6	1	0	1	1	1	1	1		0	0
7	1	1	0	1	1	1	1		0	0
									0	0
n-1				1	1	1	1	1	1	0
n				1	1	1	1	1	1	1

A 2^3 address to (8-1) channels transcoding is made for example.

Table 2. Transcoding truth table for 2^3 address to (8-1) channels

D				Ch 1	Ch 2	Ch 3	Ch 4	Ch 5	Ch 6	Ch 7
1	0	0	0	0	0	0	0	0	0	0
2	0	0	1	1	0	0	0	0	0	0
3	0	1	0	1	1	0	0	0	0	0
4	0	1	1	1	1	1	0	0	0	0
5	1	0	0	1	1	1	1	0	0	0
6	1	0	1	1	1	1	1	1	0	0
7	1	1	0	1	1	1	1	1	1	0
8	1	1	1	1	1	1	1	1	1	1

In order to achieve the input energy level necessary to activate each of the electro hydraulic executors, signals coming from transcoder need to be amplified. This function is accomplished by the heptachannel amplifier shown in Figure 7.

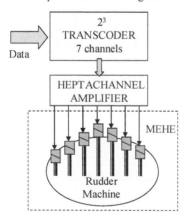

Figure 7. Heptachannel subsystem

3 INCREASING THE SAFETY

3.1 *Preventing the premature usage of the most activated electro-hydraulic devices*

Analyzing the system diagram drawn in figure 5 and seeing the tables 1 and 2, it is obviously that the most activated are executors S_1 and P_1. To avoid the long time operation in this regime, the role of prime must be cherish equally.

For the above mentioned purpose the author suggest two solutions:
– step or multistep translating the inputs of channels
– random distribution

3.1.1 *Step or multistep translating the inputs of channels*

The technique consists in shifting the amplifiers inputs to change the activation frequency of the electro-hydraulic executors. Switching is controlled using a counter of actuations imposed by the operator, as Figure 8 and Table 3 are showing.

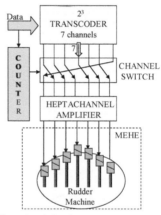

Figure 8. Channel switch subsystem

The switch is shifting the channels one time after a selected number of actuations.

Table 3. Switching truth table perm#1

D				Ch 1	Ch 2	Ch 3	Ch 4	Ch 5	Ch 6	Ch 7
1	0	0	0	0	0	0	0	0	0	0
2	0	0	1	0	1	0	0	0	0	0
3	0	1	0	0	1	1	0	0	0	0
4	0	1	1	0	1	1	1	0	0	0
5	1	0	0	0	1	1	1	1	0	0
6	1	0	1	0	1	1	1	1	1	0
7	1	1	0	0	1	1	1	1	1	1
8	1	1	1	1	1	1	1	1	1	1

3.1.2 *Random distribution*

The technique consists in switching the amplifiers inputs randomly as in the Table 4.

Table 4. Random switching truth table

D				Ch 1	Ch 2	Ch 3	Ch 4	Ch 5	Ch 6	Ch 7
1	0	0	0	0	0	0	0	0	0	0
2	0	0	1	0	1	0	0	0	0	0
3	0	1	0	0	1	0	0	0	1	0
4	0	1	1	0	1	0	1	0	1	0
5	1	0	0	0	1	0	1	0	1	1
6	1	0	1	0	1	1	1	0	1	1
7	1	1	0	0	1	1	1	1	1	1
8	1	1	1	1	1	1	1	1	1	1

3.2 *Increasing the safety of steering and navigation*

By placing pumps for each correspondently pair of electro-executors the safety problem is solved, as figure 9 and figure 10 are showing.

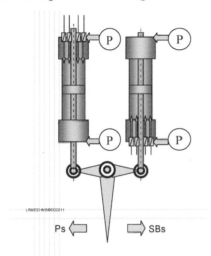

Ps ⇐   ⇒ SBs

Figure 9. Placing pumps for each board executor

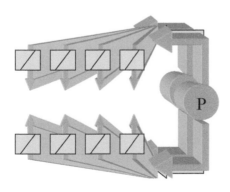

Figure 10. Placing pumps for each board correspondent pair of executors

4 THE ENTIRE STEERING SYSTEM

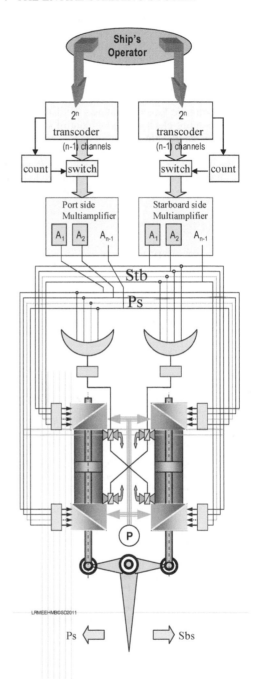

Ps ⇐   ⇒ Sbs

Figure 11. Safetier digital multichannel electro-hydraulic steering system

161

5 CONCLUSIONS

The here-by presented solution combines the advantages of the two above-mentioned valves within a compartment of an electro hydraulic executor, being (cap)able to be digitally activated with a very good transfer of energy. The Multichannel Electro Hydraulic Executor (MEHE[2]) is a discrete/proportional device. The Digital Multichannel Electro Hydraulic Execution allows the data transfer within the entire electric steering control subsystem providing a safely ship's steering and increasing the navigation security. The multiway admission makes possible the rudder machine operation even if one or more channels become defective. Using individual pumps on each way the steering safety is increasing.

REFERENCES

Dordea, Şt. 1998. Ph.D. lecture, *Sisteme de guvernare cu acţiune continuă şi comandă digitală*, Universitatea "Dunărea de Jos", Galaţi, Romania.

Dordea, Şt. 1998. Ph.D. lecture, *Contribuţii la transferul de energie în sistemele electrohidraulice de guvernare navale*, Universitatea "Dunărea de Jos", Galaţi, Romania.

Dordea, Şt. 2002. Ph.D. thesis, *Transferul de energie în sistemele electrohidraulice de guvernare navale*, Universitatea "Dunărea de Jos", Galaţi, Romania.

Dordea, Şt. 2010. Proportional Steering System, *Analele Universităţii Maritime Constanţa*, ISSN 1582-3601, vol. XIII: 165-173. Constanţa, Romania.

Dordea, Şt., Zburlea, E. 2010. Anti-Hunt Control System, *Analele Universităţii Maritime Constanţa*, ISSN 1582-3601, vol. XIII: 173-179. Constanţa, Romania.

Dordea, Şt., Zburlea, E., Badea, M. 2009. Optoelectronic Hydraulic Device, *Analele Universităţii Maritime Constanţa*, ISSN 1582-3601, vol. XII: 309-317, Constanţa, Romania.

Dordea, Şt., Zburlea, E., Badea, M. 2009. Naval Steering Control By Means Of Optic Tracks, *Analele Universităţii Maritime Constanţa*, ISSN 1582-3601, vol. XII: 317-323, Constanţa, Romania.

Dordea, Şt. 2008. Monitoring the steering failures, *Ovidius University Annals of Mechanical, Industrial and Maritime Engineering*, Tom I, 2008, cod CNCSIS 603, Editura Ovidius University Press, ISSN 1223-7221, Vol. X, Constanţa, Romania.

Dordea, Şt. 2007. Double Track Steering Failure Monitoring System, *Annals of the „Dunarea de Jos" University of Galati, Fascicle III – Electrotechnics, Electronics, Automatic Control, Informatics*, cod CNCSIS 482 ISSN 1221-454X, „Dunărea de Jos" University of Galati, Romania.

Dordea, Şt. 2007. A Ship's Rate of Turn Steering Control, *A XXXII-a Sesiune de comunicări ştiinţifice cu participare internaţională „TEHNOLOGII MODERNE ÎN SECOLUL XXI, Academia Tehnică Militară*, Bucureşti, Romania.

Dordea, Şt. 2006. Maritime automatic steering system with off-course monitor, *Analele Universităţii Maritime Constanţa*, ISSN 1582-3601, vol. IX: 227-230, Constanţa, Romania.

Dordea, Şt. 2006. Versatile automatic steering system with off-course monitor, *Analele Universităţii Maritime Constanţa*, ISSN 1582-3601, vol. IX: 231-234, Constanţa, Romania.

Dordea, Şt. 2005. Digitaly activated electro hydraulic distributor, *A XXX-a Sesiune de comunicări ştiinţifice cu participare internaţională „Tehnologii moderne în secolul XXI", Academia Tehnică Militară, Bucureşti, 3-4 noiembrie 2005*, ISBN 973-640-074-3, pp. 13.34-13.37, 4 pgs, Constanţa, Romania.

Dordea, Şt. 2004. Ship's rate-of-turn steering control, *Analele Universităţii Maritime Constanţa*, ISSN 1582-3601, vol. VII: 197-202, Constanţa, Romania.

---

[2] MEHE is a concept developed since 1998 by Dr Stefan Dordea (see the 2002 doctoral thesis *"The energy transfer within naval electro-hydraulic steering systems"*)

Safe Shipping and Environment in the Baltic Sea Region

22. Towards the Model of Traffic Flow on the Southern Baltic Based on Statistical Data

A. Puszcz & L. Gucma
Maritime University of Szczecin, Poland

ABSTRACT: The paper presents methods and models used for analysis of ships traffic in Southern Baltic on chosen areas. Ships' traffic has been analyzed by means of statistical methods with use of historical AIS data obtained from HELCOM. The paper presents probabilistic models of ships' traffic spatial distribution and its parameters. The results could be used for safety and risk analysis models in given area and for creation of general models of ships' traffic flows.

1 INTRODUCTION

Ships' traffic is the most important factor of ships safety. A good understanding of traffic stream behavior is necessary for an efficient design of traffic facilities. One needs to understand how vessels interact with each other and with the traffic facility in order to design such facilities better. That is, one needs to understand the stream behavior under various traffic conditions.

The stream behavior is complex because the traffic stream is an outcome of human maneuvering process. The precise knowledge of ships' traffic phenomenon and its processes is significant for navigational safety analysis. AIS (Automatic Identification System) gives great opportunity for traffic monitoring but also for discovering basic processes which rules the traffic of ships in confined areas. Results of statistical analysis present in this paper conducted to find ships' positions spatial distribution and its parameters of ships' traffic in analyzed area.

One of the main problems posed by marine traffic engineering is to determine the optimum parameters of new constructed or modernized parts of the waterways. Depending on the type of waterway parameters that can be obtained are for example, lane width or diameter of the turning circle. These parameters are usually determined in one of two methods: a cheaper but less accurate analytical method or the more expensive and more accurate simulation method.

This paper presents how utilize historical AIS data to explore the existing maneuver pattern in the area, estimate parameters describing traffic flow and used for the construction of a new method of determining a ship maneuvering area. Paper presents first steps leading to create a general model of traffic flow- statistical analysis of ships traffic flow on selected areas of Southern Baltic.

2 ANALYZED AREA

2.1 *General information*

The Baltic Sea has relatively dense traffic. One of the most visited ports in southern Baltic is Świnoujście. The port had more than 8,8 million tons of throughput in 2008, and a terminals for coal, chemicals and raw materials. The number of calls in 2008 is about 7000. It also has links for passengers, cars, trains and ferry services to Germany. Świnoujście is one of the most visited ports in Poland.

There are two main flows to Świnoujście from Arkona. The daily average of ships' passages using these lanes (going to Świnoujście) is around 25 including ferries, passenger vessels and fishing boats. The number of ships passing through the area is expected to grow significantly over the years to come due to the general increase of trade, building of new LNG Terminal in Świnoujście and the growing economy of the eastern Baltic countries.

Due to comprehensive researches paper presents results only for the approach to Świnoujście on the section form buoys 5-6 to head of breakwater.

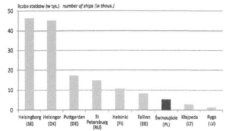

Figure. 1 Ship calls top selected sea ports in the Baltic Sea Region in 2008 [source: GUS]

Figure 2. Analyzed area- approach to Świnoujście.

2.2 Navigational conditions

Świnoujście Roads, the inner part of Pomorska Bay, is approached from the N or E channels, passing respectively, W or S of Oder Bank (54°21'N 014°25'E), known locally as Odrzana Bank; the shoal area, which is extensive with a least depth of 4.8m, white sand, is fairly steep-to, except on its N side. The navigation channel is 32nm long with a varying width of 180-200m and a depth of 14.3m.

The port entrance is protected by breakwaters. Weather conditions are those normally encountered in temperate zone. Strong NE winds cause heavy seas in the bay. The port is generally ice free all year, however difficulty maybe experienced between the beginning of January and the middle of March.

3 STATISTICAL MODEL OF SHIPS TRAFFIC STREAMS

3.1 Spatial distribution of ships traffic

The theory of traffic flow of ships involved in the phenomena described the movement of many vessels through the traffic lane in the some chosen period of time. One of the main parameters describing the traffic flow is the distribution, describing the ship's hull position in relative to the axis of the track.

The information about the position of the vessel's center of gravity, the shape of the waterline and the course are used to define the distribution. A simple approach to describe traffic streams is their characterization by means of a single, specific resolution. It should be noted that there are different types of distribution in relation to the track section: straight track or bend. The most common and used distributions are as follows:

– normal distribution with PDF (probability density function):

$$d_l(y) = \frac{1}{\sigma_l\sqrt{2\pi}} e^{-\frac{(y-m_l)^2}{2\sigma_l^2}}$$

(1)

– logarithmic distribution:

$$d_l(y) = \frac{1}{y \cdot \sigma_l\sqrt{2\pi}} e^{-\frac{(\ln y - m_l)^2}{2\sigma_l^2}}$$

(2)

– gamma distribution:

$$d_l(y) = \frac{a^p}{\Gamma(p)} \cdot y^{p-1} \cdot e^{-a \cdot y}$$

(3)

– logistic distribution:

$$d_l(y) = \frac{z}{\beta(1+z)^2}, \text{ where } z = e^{\frac{v-\alpha}{\beta}}$$

(4)

where:
y – distance to the axis, m – average of ships distance to the waterway axis, σ – standard deviation of ships distance to the waterway axis, $\Gamma(p)$ – gamma function, $\beta(p, q)$ – beta function, a, p, q, β, α – other parameters of distributions.

The width of traffic flow is fundamental importance in its assessment. In order to describe these traffic streams is necessary to determine the characteristics of the distributions of the width of the traffic lane.

3.2 Time distribution of ships traffic

The aim of the study is to determine a statistical model that can describe the flow ships and define the parameters obtained models depending on time of day or year. The results obtained can be used to their practical application to the simulation models for traffic flow parameters, and may also facilitate the prediction of the expected values of the stream traffic on the studied region in the forthcoming years.

The basic input parameter of traffic streams for modeling vessels behavior is also the intensity. Vessel traffic along the fairway is a process that is influenced by many factors change over time and the length of the track. They make the movement becomes a random process, and his description of the probabilistic models are used. The flow of ships on the seaway can be represented on the timeline, where the transition moments of ships through the center are random events. Random stream of vessels can be analyzed by examining: distribution of number of vessels passing through this point in time ΔT; the distribution of time intervals between the vessels; the distribution of local speed vessels.

The number of ships passing given point of the waterway in case when vessels have freedom of selecting speed and manoeuvre can be described as Poisson distributed stochastic process where probability of appearance of $X=n$ ships in Δt time is:

$$P(X=n)=\frac{(\lambda\Delta t)^{n}}{n!}e^{-(\lambda\Delta t)} \text{ for } \Delta t \geq 0 \qquad (5)$$

where:
λ-traffic intensity [ships/s]
Probability, that in Δt time no ship will appear is:

$$P(X=0)=e^{-(\lambda\Delta t)} \text{ for } \Delta t \geq 0 \qquad (6)$$

4 RESEARCH METHOD

Determine of the exact vessels path and their corresponding speed allows describing traffic stream. The results should be formulated in such a way, that they are general applicable and can be used as input for a maritime model. To reliably describe a vessel's path and speed in a model the following should be determined:

1 Describe the spatial distribution of vessels in a certain section;
2 Describe the lateral vessel speed distribution of vessels in a certain section;
3 Describe the vessel speed distribution on a certain location;
4 Take into account that the 3 distributions mentioned above depend on:
   – Vessel type;
   – Vessel size;
   – Vessel heading / destination;
   – Type of waterway segment (straight / bend)
   – Width of the (for that specific vessel type and size) navigable waterway;
   – Wind speed and wind direction;
   – Current speed and current direction;
   – Visibility;
   – Other external influences;
   – Interaction with other vessels.
5 Describe the mutual dependence between two spatial successive distributions (to connect the different sections, in order to assemble an individual vessel path and correct speed development)

The main factors affecting the movement of vessels in relation to the axis of the traffic lane are the size of the vessels, meteorological conditions (waves, wind), the experience of the officer. An additional parameter is the one characterizing the intensity of shipping traffic. It depends largely on the economic situation in the market and current season.

The research consisted mainly of matching the distribution of traffic in relation to the axis of the traffic lane on Świnoujście Harbour. The procedure for determining the type of distribution of selected parameters is as follows:
– Information on traffic received from the AIS (for the period 01.2008–02.2009) is used.
– Data is divided on the season, day or night time and type of traffic (inbound or outbound).
– Grouped samples studied separately as separate random variables.
– Defined a random variable as the location of the vessel in relation to the axis of the track.

The data obtained was transformed by means of the middle line method used for the bends and straight sections of the fairway [3]. This method (Fig. 3) uses as a reference the center of the track segments approximated by the length of the section (i). Sections have the shape of part-circular or rectangular. Based on data about the course, waterline and geometric center of waterline are calculated coordinates of extreme points of the vessel (right and left), then their distances to the track axis (the axis of reference). Corresponding distributions have been matched by means of the distance tables [3].

The analyses have been made on a selected set of data from approach to Świnoujście. The midpoint of the navigation channel or traffic lane is used as origin. Thus, mean values are the average distance from the mid point of the navigation channel. Gates midpoint was situated in the middle of the approach channel, between buoys no 7 and no 8 in a way that the buoys 7-8 are in one line with the gate assigned. Others gates have been determined in the same way.

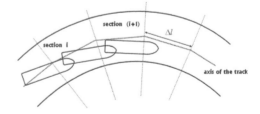

Figure 3. Method for the measure of the track, the track center line approximation to the polygon and the track division into sections [4].

4.1 Results

4.1.1 Average vessel path and speed

Figure 4 shows an example of a spatial distribution, derived from the empirical data at a certain gates. On the X-axis a value of zero correspondents to the middle of the axis. A positive value for X means that the vessel sails more to the starboard side.

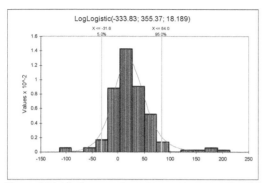

Figure. 4 Distribution over the waterway at cross section 7-8 for vessels length L<120 m, January 2009.

Their goodness of fit is first determined by performing a Chi-square test (χ_2). This test determines the degree of agreement between the empirical distribution and the theoretical distribution. The hypothesis is that there is no significant difference between those distributions. The confidence level (answering the question what is significant) is set on 95%. Also Kolmogorov- Smirinov and Anderson-Darling tests have been perform. K-S statistic, the A-D statistic does not require binning. But unlike the K-S statistic, which focuses in the middle of the distribution, the A-D statistic highlights differences between the tails of the fitted distribution and input data. In tab.1 tests results are shown.

Table. 1 Tests results for vessel approaching to Świnoujście

Test	January			
	L>120 in	L>120 out	L<120 in	L<120 out
Chi sq.	3.2500	1.1140	12.9000	15.6000
A-D	0.3841	0.2301	0.1200	0.7669
K-S	0.1084	0.1057	0.0607	0.0520
	August			
	L>120 in	L>120 out	L<120 in	L<120 out
Chi sq.	13.6700	10.4100	19.5900	24,32
A-D	2.4080	0.9426	2.3690	2.6270
K-S	0.1563	0.0908	0.0699	0.0922
	ro-ro ferry			
	Jan/IN	Jan/out	Aug/in	Aug/out
Chi sq.	14.2200	13.9600	14.2100	16.6700
A-D	0.9897	0.4087	0.9060	2.1670
K-S	0.0520	0.0465	0.0541	0.0631

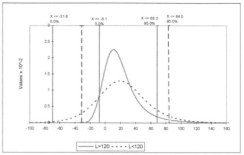

Figure.5 Distribution over the waterway at gates 7-8 for inbound vessels, January 2009.

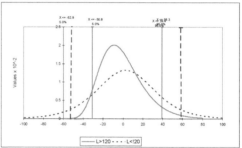

Figure. 6 Distribution over the waterway at gates 7-8 for outbound vessels, January 2009.

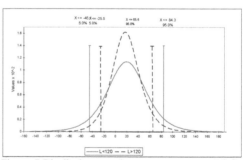

Figure. 7 Distribution over the waterway at gates 7-8 for inbound vessels, August 2008

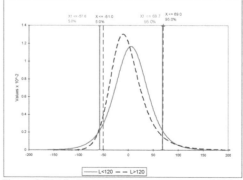

Figure. 8 Distribution over the waterway at gates 7-8 for outbound vessels, August 2008

Can be seen on the X-axis (0 is on the port side of the incoming vessels) that inbound vessels sail on the right side of waterway for winter and summer months but for outbound vessels middle of their tracks is near the middle of the channel. Analyzing mean and standard deviation of the distributions was obtained that the largest vessel (L>120 meters) sail more to the middle of the channel. The smallest size class (L<120 meters) sails more to the outer limits.

The spatial deviation over the waterway is very well approximated by a normal distribution. This is supported by calculating values for the skewness and excess kurtosis of the empirical datasets. A χ^2-test showed as well that the assumed logistic distributions are a good approximation. In 50% of obtained spatial distribution, the best describing distribution is loglogistic- and logistic-40% in other 10% of cases normal and lognormal distribution have had the best results.

Logistic distribution best matches the data around the mean value. Normal distribution is more flattened compared to the logistic distribution. There is also a difference in the description of the extreme values. Despite the good fit to the normal distribution for the value closest to the average, normal distribution adapts insufficiently to the extreme values.

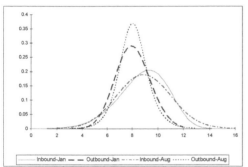

Figure.9 Vessel speed distribution on location 7-8, for all size vessels.

Although the vessels speed entering and leaving the port of Świnoujście is specified in the port regulations the graphs make clear that the vessel speed has a wide distribution. Speed mean value for inbound vessels is higher than for outbound which does not depend strongly on the seasons.

4.1.2 Intensity of ships traffic

The study focused mainly on verification of the hypothesis that the traffic flow model on the Świnoujście fairway have a Poisson distribution. The random variable defined as "the number of vessels appearing on the fairway within 4 hours."- The data was transformed so as to comply with the above-defined of random variable. Data were grouped according to season, type of traffic (inbound and outbound) and time of day, so that they formed a separate random variables, which were studied separately. This was due to the variation of the intensity of vessel traffic.

The verification was carried out using χ^2 test. Statistical tests were applied to the data according to:
- Season (winter and summer months)
- Time of day (day, night)
- Traffic (inbound, outbound)

Figures 10 to 11 are graphs match schedules for the two months of data (summer and winter) on day and night time (0800-2000 and 2000-0800) for inbound and outbound vessels. In all the verification tests performed at a level of significance $\alpha = 0.05$.

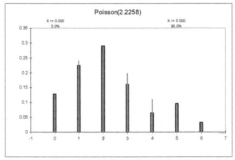

Figure. 10 Histogram with fitted distribution for gate 7-8 –inbound traffic, summer daytime.

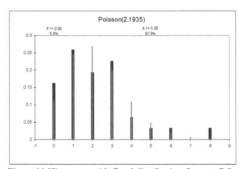

Figure.11 Histogram with fitted distribution for gate 7-8 –inbound traffic, summer, nighttime.

Tab.2 Parameters of Poisson distribution for separate random variables

Month	hrs	Inbound traffic		
		Intensity	Test Value	P Value
January	0800-2000	1.806	1.279	0.8650
	2000-0800	1.613	1.476	0.6878
July	0800-2000	2.226	0.290	0.9619
	2000-0800	2.193	1.189	0.7557
		Outbound traffic		
		Intensity	Test Value	P Value
January	0800-2000	1.742	3.364	0.4988
	2000-0800	1.484	0.634	0.8887
July	0800-2000	2.806	2.229	0.6937
	2000-0800	2.000	1.281	0.7336

The results presented in Tab.2 show that Poisson distribution could be the model of ships traffic process.

5 CONCLUSIONS

The methodology presented in the paper is the basis for the development of a mathematical model of traffic. Creation of precise model that presents results should be verified by studying on traffic flows and its distributions on different water areas. It was proved that Poisson distribution could be used for traffic modeling in analysed area.

Using this approach it is possible to obtain generic rules that describe the vessel path and vessel speed in many different areas. To do so, the case study area (Southern Baltic) will be split into several characteristic waterway and segments and the location specific results will be generalized to their specific segments.

Can be concluded that by using an analysis of historical AIS data, clearly more insight is obtained in the detailed individual vessel behavior. This understanding of the behavior can be formulated in generic rules. These rules can be implemented in maritime model, which improves the simulation of the individual vessel path and vessel speed.

It should be noted that the distribution applies solely to the AIS registered ship traffic. Thus, the leisure boats and fishing boats will be modeled separately to form a complete ship traffic distribution.

Leisure boat traffic is only to a limited extent present in winter, spring and fall. Thus, seasonal variations will be included in the model to account for pleasure boat traffic in the total ship traffic volume and in the ship traffic distributions.

Further work should be focused on improving the model that will involve the examination of traffic flows on the open and limited water, depending on the hydro meteorological conditions, ship traffic, and depending on the existing dangers (e.g. shoal).

ACKNOWLEDGEMENT

This paper was created with support of EfficienSea project; partially EU founded Baltic Sea Region Programme 2007-2013.

REFERENCES

[1] IMO NAV 51/3/X March 2005 SUB-COMMITTEE ON SAFETY OF NAVIGATION 51st session. Overview of the ships' traffic in the Baltic Sea – HELCOM 2008.
[2] Overview of the ships' traffic in the Baltic Sea. HELCOM 2008.
[3] Guziewicz J. Ślączka W., Methods for determining the maneuvering area of the vessel used in navigating simulation studies", VII MTE Conference, Szczecin 1997.
[4] Gucma L. "The study of probabilistic characteristics of traffic flow on the fairway Szczecin-Świnoujście", Szczecin 2005.

23. Incidents Analysis on the Basis of Traffic Monitoring Data in Pomeranian Bay

L. Gucma & K. Marcjan
Maritime University of Szczecin, Poland

ABSTRACT: In this paper preliminary analysis of grounding incidents was presented for the future use in navigational safety management system development. The analysis is focused on the area of Pomeranian Bay. Grounding incidents model is created with use of AIS data. The distance of vessel from the dangerous depth and draught/depth ratio was considered as the main factors of navigational incident in presented model.

1 INTRODUCTION

Marine transport development causes the increase of the intensity of ships traffic and ships dimensions. The navigational risk increase requires more sophisticated methods of its assessment. Marine accidents such as grounding are very rare and serious safety analysis should also include incident (near misses) analysis. The incident analysis is very important factor in navigational safety analysis due to the incidents are usually not reported. Probability of the ships grounding is one of the most important factors influencing the navigational safety [2]. Based on the data from HELCOM in 2008, there were 60 reported groundings in whole Baltic Sea, and 32 of them were reported in the south western Baltic Sea. As it is shown on Figures 1 and 2, in 2009 the number of reported groundings decreases to 38, but still about half of them were reported in south-western Baltic Sea.

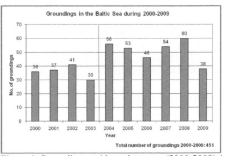

Figure 1 Grounding accidents between (2000-2009) in south-western Baltic Sea

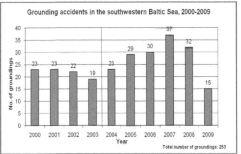

Figure 2 Grounding accidents between (2000-2009) in Baltic Sea

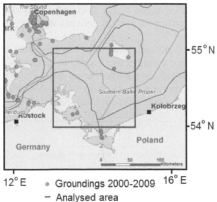

Figure 3 Ships groundings on southern Baltic, analysed area [4]

Presented in this paper analysis of grounding incidents is devoted to the Pomeranian Bay area exactly the area between 13^0 00'E - 15^0 00'E and 54^0

00'N - 55⁰ 00'N. This area extends between Bornholm, Rügen and the approach to Świnoujście. It may be considered as costal area, where some places have relatively low depth. Those are depicted in Figure 4. Regions with depth of 10 m or less extends around the Rügen Island, around Bornholm close to Polish and German shore and also north of approach to Świnoujście.

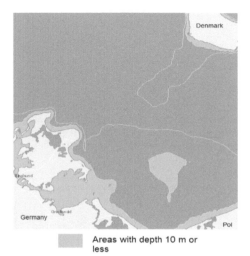

Figure 4 Map of selected area with isobaths.[5]

2 MODEL DETERMINING THE GROUNDING INCIDENTS

Model used for determining the grounding incidents on selected area. The model is based on AIS data, from where static and dynamic data allow one to determine the information on the ship and its positions. Application written in C# [3] consists of three sections:

1 First section decodes data retrieved from the AIS and records the routes of vessels navigating within the analysed area. Data is segregated and written to the appropriate database tables. Vessels of 6m depth or more are taken into account.

Date	MMSI	NavStat	Latitude	Longitude	Sog	Cog	Heading
31 22:58:2	265276...	0	3259,320	837,912	9,8	313,4	315,0
2008-01-...	308079...	0	3250,800	847,525	10,3	333,0	334,0
2008-01-...	255801...	0	3255,170	851,210	9,8	302,0	304,0
2008-01-...	235642...	0	3277,290	889,569	17,5	80,0	82,0
2008-01-...	261392...	0	3283,820	804,886	8,3	255,9	255,0

Figure 5 Database table with AIS dynamic data.

2 Second section examines individual positions of the vessel in terms of distance from depths. The depths of less than 140% of vessel's draught are taken into account. Then the lowest depth, and the smallest distance to this depths is recorded.

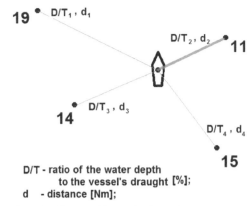

D/T - ratio of the water depth to the vessel's draught [%];
d - distance [Nm];

Figure 6 A diagram of lowest depth and distance selection.

3 Third section is a model of positions extrapolation. On the basis of dynamic data this part of programme is searching for the previous position $(\varphi(t0), \lambda(t0))$ and following position $(\varphi(tn), \lambda(tn))$ of the vessel in vicinity of the dangerous depth. Both positions are taken into account only if the time difference between them and the reference position $(\varphi(tr), \lambda(tr))$ is less than 6 minutes. Extrapolation algorithm calculates every second position and the distance from the depth between the previous, reference and the following position. The result is a position of a vessel that is the closest to the smallest depth in vicinity of the vessel.

$$\lambda_i = \lambda_0 + \sum_{j=0}^{i} \sin(COG_j) \cdot V_j + i \cdot \frac{\lambda_r - \sum_{k=0}^{r} \sin(COG_k) \cdot V_k}{r} \quad (1)$$

$$\varphi_i = \varphi_0 + \sum_{j=0}^{i} \cos(COG_j) \cdot V_j + i \cdot \frac{\varphi_r - \sum_{k=0}^{r} \cos(COG_k) \cdot V_k}{r} \quad (2)$$

where:

φ_i – latitude in time t_i between previous position and reference position of the vessel,

λ_i – longitude of position number i between previous position and reference position of the vessel,

COG_j – Course Over Ground in time t_j,

V_j - speed of the vessel in time t_j,

r – number of extrapolated positions between previous position and reference position of the vessel.

$$COG_j = COG_0 + (t_j - t_0) \cdot \frac{dCOG}{dt} \quad (3)$$

$$V_j = V_0 + (t_j - t_0) \cdot \frac{dV}{dt} \quad (4)$$

3 RESULTS

The result of the analysis of grounding incidents on the area between $13^0\,00$'E - $15^0\,00$'E and $54^0\,00$'N - $55^0\,00$'N within 6 month time period (1th January 2008 – 30th June 2008) were 230 grounding incidents calculated by the algorithm. Figure 7 is presenting analysed area with marked vessel position divided due to the ratio D/T. The 10m and 12m isobaths are shown, to mark the places with low depth. All the ports and their surroundings were excluded from the examined area.

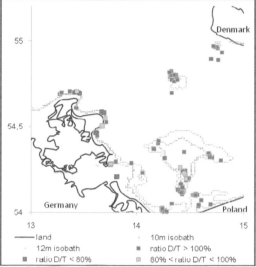

Figure7 Grounding incidents on selected area from 01.01.2008 to 30.06.2008

The ratio of vessel draught to the depth of the water is shown in figure 8. 42 vessels were passing close to the depth which was lower than their draught. The other 188 vessel were close to the depth which might have been a problem to pass it safe.

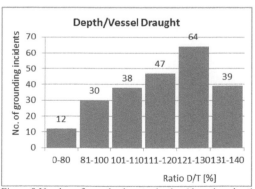

Figure 8 Number of vessels close to depth with assigned ratio D/T

Generally there were 3 ships that were passing the dangerous depth with distance of 2 cables or less. It must be remembered that the distance is measured between the calculated depth and the position of vessel's antenna. The actual distance from the depth could have been lower than 2 cables. All the distances of the ship to the depth are depicted in figure 9. According to experts the distance of 0,7 Nm can be interpreted as a safe distance, but taking into account the conditions of open water, getting so close to a dangerous depth contours can be regarded as an unjustified risk situation or a grounding incident.

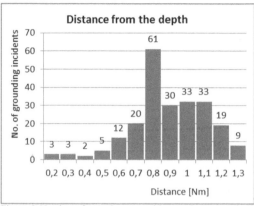

Figure 9 Number of vessels approaching to dangerous depth with assigned distance.

Length of most of the vessels which had approached close to dangerous depths is between 101m and 160m. Most vessels length range extends between 101m and 120m, these are vessels of draught (5-7m). Such depth are mainly found in coastal areas. It means it is very likely that some surroundings of ports aren't sufficiently cut off.

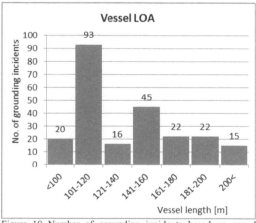

Figure 10 Number of grounding incidents based on vessel length.

4 CONCLUSIONS

Presented results of grounding incidents constitute a valuable source of information about the areas with low water depth around which vessels are passing with dangerous distances. The incident places obtained by presented model are very close to the grounding accidents on analyzed area (Fig.3). Obtained results will be used in navigational safety management system development, which is described in [1]. There is still some work to be done to verify models results but partially verification of the results presented, is overlapping positions of calculated grounding incidents to real groundings, in the examined area.

REFERENCES

Gucma L., Marcjan K. 2011. The Incident Based System of Navigational Safety Management of on Coastal Areas. Proceedings of the 8th International Probabilistic Workshop. Szczecin.

Przywarty M. 2008. Models of Ships Groundings on Coastal Areas. Journal of Konbin Vo-lume 5, Number 2 / 2008. Versita, Warsaw.

ECMA International. 2006. C# Language Specification (4th ed.).

HELCOM. 2009. Report on shipping accidents in the Baltic Sea area during 2009.

http://maps.helcom.fi/website/mapservice/index.html

http://www.io-warnemuende.de/topography-of-the-baltic sea.html

24. Model of Time Differences Between Schedule and Actual Time of Departure of Sea Ferries in the Świnoujście Harbour

L. Gucma & M. Przywarty
Maritime University of Szczecin, Poland

ABSTRACT: The paper presents the assumptions of model of time differences between schedule and actual time of departure of sea ferries in the Świnoujście Harbour. Data necessary to build the model were collected during the measurements in the port. The model was used for the construction of a comprehensive model of vessels traffic in the south Baltic Sea area. In the next stage the simulation experiment was carried out. The verification of the developed model was conducted and the positions of simulated sea accidents and the encounter situations were established. The output of the simulation can be used to assess the navigational safety.

1 INTRODUCTION

Simulation methods are nowadays the most common approach to the navigational safety assessment. The present-day systems, due to their complexity, require construction of dedicated and suitably chosen methods. Generalized empirical method, using the Monte Carlo simulation and theory of ships traffic flows gives the most satisfactory results in this respect (Gucma, 2005). In most cases, the intensity of traffic can be described by a Poisson distribution (Gucma, 2004). However, if the movement of ships takes place in a predetermined schedule Poisson distribution cannot be used. The goal of this paper is to present developed model of the time differences between actual departures and the schedule of sea ferries. Developed model will allow to simulate scheduled traffic. To achieve the assumed goal the actual differences between time of departure and schedule of sea ferries in the Świnoujście Harbour were measured. Next the statistical analysis of the results was performed and the theoretical distribution was fitted, this allows to implement the built model to a microscopic model of traffic.

2 RESEARCH DESCRIPTION

The research was performed since 22.07.2010 to 04.08.2010 in The Ferry Terminal in Świnoujście Harbour (Figure 1). The times of departure were measured, next the differences between time of departure and scheduled time were calculated.

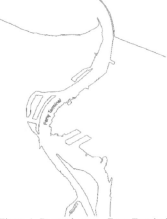

Figure 1. Research area – Ferry Terminal in The Świnoujście Harbour.

$$d_t = t_a - t_s \tag{1}$$

where:
d_t - difference between actual and scheduled time of departure,
t_a – actual time of departure,
t_s – scheduled time of departure

About 60 measurements were performed during research period.

The measurements were carried out for almost all ferries entering to the Świnoujście Harbour. The list of ferries is presented below:
- M/F Polonia;
- M/F Kopernik;
- M/F Pomerania;

- M/F Wawel;
- M/F Skania;
- M/F Gryf;
- M/F Galileusz;
- M/F Wolin.

3 RESULT ANALYSIS

The following results were obtained:
- Mean difference between actual and scheduled time of departure: 4min;
- Maximum difference between actual and scheduled time of departure: 34min;
- Minimum difference between actual and scheduled time of departure: -9min (ferry unberthed before scheduled time);
- Standard deviation: 8.2min;
- Skewness: 1.84;
- Mode: 3min.

Histogram of the differences between actual and scheduled time of departure is presented in Figure 2.

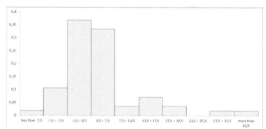

Figure 2. Histogram of the differences between actual and scheduled time of departure of Ferries in the Świnoujście Harbour.

According to characteristics of a random variable, calculated parameters and on the basis of the histogram following theoretical distributions were chosen for further analysis.

Erlang Distribution:
- Parameters:
  - m – shape parameter (positive integer)
  - β – continuous scale parameter (β>0)
  - γ – continuous location parameter
    ($\gamma \equiv 0$ yields the two-parameter Erlang distribution)
- Domain:
  - $\gamma \leq x < +\infty$
- Probability Density Function:
  - $f(x) = \dfrac{(x-\gamma)^{m-1}}{\beta^m (m-1)!} \exp(-(x - \gamma)/\beta)$

Lognormal Distribution:
- Parameters:
  - σ – continuous parameter (σ>0)
  - μ – continuous parameter

- γ – continuous location parameter
  ($\gamma \equiv 0$ yields the two-parameter Lognormal distribution)
- Domain:
  - $\gamma < x < +\infty$
- Probability Density Function:

$$f(x) = \frac{\exp\left(-\frac{1}{2}\left(\frac{\ln(x-\gamma)-\mu}{\sigma}\right)^2\right)}{(x-\gamma)\sigma\sqrt{2\pi}}$$

Logistic Distribution:
- Parameters:
  - σ – continuous scale parameter (σ>0)
  - μ – continuous location parameter
- Domain:
  - $-\infty < x < +\infty$
- Probability Density Function:

  - $f(x) = \dfrac{\exp(-z)}{\sigma(1 + \exp(-z))^2}$

  where $z \equiv \dfrac{x - \mu}{\sigma}$

Log-Logistic Distribution:

- Parameters:
  - α – continuous shape parameter (α>0)
  - β – continuous scale parameter (β>0)
  - γ – continuous location parameter
    ($\gamma \equiv 0$ yields the two-parameter Log-Logistic distribution)
- Domain:
  - $\gamma \leq x < +\infty$
- Probability Density Function:

  - $f(x) = \dfrac{\alpha}{\beta}\left(\dfrac{x-\gamma}{\beta}\right)^{\alpha-1}\left(1 + \left(\dfrac{x-\gamma}{\beta}\right)^{\alpha}\right)^{-2}$

Generalized Extreme Value Distribution:

- Parameters:
  - k – continuous shape parameter
  - σ – continuous scale parameter
    (σ> 0)
  - μ – continuous location parameter
- Domain:
  - $1 - k\dfrac{x-\mu}{\sigma} > 0$ for $k \neq 0$
  - $-\infty < x < +\infty$ for $k = 0$
- Probability Density Function:

  - $f(x) = \begin{cases} \frac{1}{\sigma}\exp\left(-(1-kz)^{1/k}\right)(1-kz)^{1/k-1} & \text{for } k \neq 0 \\ \frac{1}{\sigma}\exp\left(-z - \exp(-z)\right) & \text{for } k = 0 \end{cases}$

  where $z \equiv \frac{x-\mu}{\sigma}$

Due to domain restriction to the positive numbers concerning some of the chosen distribution the following modification of time differences was performed:

$$d_m = d_r - d_{t\,min} \qquad (2)$$

where:

d_m - difference between actual and scheduled time of departure used to distribution fitting;

d_r- difference between actual and scheduled time of departure;

$d_{t\,min}$- lowest difference between actual and scheduled time of departure (-9min).

Parameters of fitted distribution are presented in the Table 1. Histograms of the time differences and the fitted distributions are presented in Figure 3.

Table 1. Parameters of fitted distributions

	Distribution				
	Erlang	Lognormal	Logistic	Log-Logistic	Generalized Extreme Value
Parameters	$m = 3$ $\beta = 3.8$ $\gamma = 0$	$\sigma = 0.5$ $\mu = 2.4$ $\gamma = 0$	$\sigma = 4.5$ $\mu = 13$	$\alpha = 3.4$ $\beta = 11.3$ $\gamma = 0$	$k = 0.2$ $\sigma = 4.7$ $\mu = 9.1$

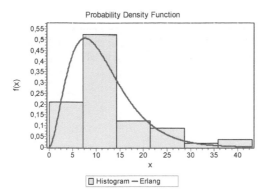

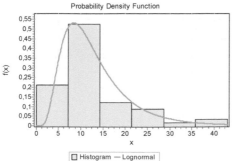

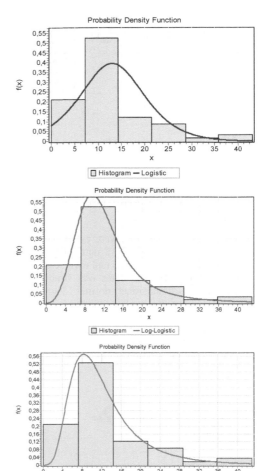

Figure 3. Histograms of the time differences and the probability density functions of fitted distributions.

Analysis of fit goodness was conducted on the basis of Kolmogorov-Smirnov Test and the Anderson-Darling Test. For both tests Generalized Extreme Value Distribution was best fitted theoretical distribution.

4 APPLICATION OF THE MODEL TO NAVIGATIONAL SAFETY ASSESSMENT

The chosen distribution was next used to simulate sea ferries traffic on the routes to and from Świnoujście Harbour. This traffic is a part of the microscopic model of ships traffic which is a module of navigational safety assessment model. Diagram of fully developed stochastic model of navigational safety assessment model is presented in Figure 4 (Gucma&Przywarty, 2007).

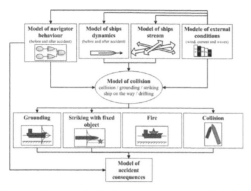

Figure 4. Diagram of fully developed stochastic model of navigational safety assessment.

Coordinates of waypoints were estimated on the basis of sea chart analysis, AIS data from Helcom database (Helcom, 2006, 2007) and own seamanship experience. Variability of routes was simulated by the use of two-dimensional normal distributions (Figure 5). Layout of simulated routes is presented in Figure 6. Navigational safety assessment model was implemented in Delphi language. Interface of model, which allows to enter initial data and to simulation supervision, is presented in Figure 7.

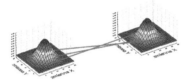

Figure 5. Distributions of waypoints coordinates.

Figure 6. Simulated routes.

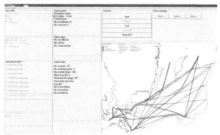

Figure 7. Model interface.

Model of navigational safety assessment allows to establish the coordinates of simulated positions of collisions, groundings and fires it also allows to simplified assessment of sea accidents consequences (Gucma&Przywarty, 2007).

In order to verify the developed model of ferry traffic simulation experiment was conducted. On the basis of the experiment it can be stated that developed model of time differences between schedule and actual time of departure faithfully reflects the traffic of sea ferries in Świnoujście Harbour. Because sea accidents involving sea ferries are very rare, it was decided to study the position of encounter situations. The results of this experiment are shown in Figure 8.

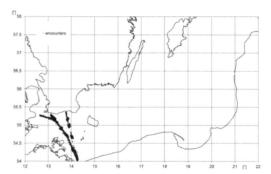

Figure 8. Positions of simulated encounter situations involving sea ferries on routes to and from Świnoujście Harbour.

Additionally, during the simulation positions of simulated groundings, collisions and fires were established. These positions are shown in Figure 8. Particular results are presented in table 2.

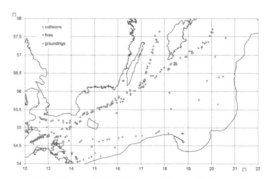

Figure 9. Positions of simulated groundings collisions and fires (10 times 5 years of simulation).

Table 2. Particular results of simulation experiment.

Type of accident	Number of accidents	Simulated time	Mean number of accidents per year	Mean time between accidents
Collision	133	10 x 5 years	2,66	0,38 years
Fire	126	10 x 5 years	2,52	0,40 years
Grounding	25	10 x 5 years	0,5	2 years
Encounter situation involving ferry	6734	1 year	6734	1.3h

5 CONCLUSIONS

Developed model of time differences between schedule and actual time of departure of sea ferries in the Świnoujście Harbour allows to simulate scheduled traffic of ships. It was proved that it can be used to simulation of traffic of sea ferries navigating on routes to and from Świnoujście Harbour. To confirm the possibility of its use in other ports, further investigation is needed.

ACKNOWLEDGEMENT

This paper was created with support of EfficienSea project; partially EU founded Baltic Sea Region Programme 2007-2013.

REFERENCES

Gucma L. 2004. Badanie probabilistycznych charakterystyk strumienia ruchu statków na torze wodnym Szczecin – Świnoujście. Zeszyty Naukowe nr 75, AM Szczecin

Gucma, L. 2005. Risk Modelling of Ship Collisions Factors with Fixed Port and Offshore Structures. Szczecin: Maritime University of Szczecin.

Gucma L.Przywarty M. 2007 Probabilistic method of ships navigational safety assessment on large sea areas with consideration of oil spills possibility International Probabilistic Symposium. Ghent.

Gucma L.Przywarty M. 2007 The model of oil spills due to ships collisions in Southern Baltic area Proc. of Trans-Nav Conference. Gdynia.

Helcom. Report on shipping accidents in the Baltic Sea area for the year 2005. Helsinki Commission Baltic Marine Environment Protection Commission (HELCOM). Draft. 2006.

Helcom. Report on shipping accidents in the Baltic Sea area for the year 2006. Helsinki Commission Baltic Marine Environment Protection Commission (HELCOM). Draft. 2007.

25. Simplified Risk Analysis of Tanker Collisions in the Gulf of Finland

F. Goerlandt, M. Hänninen, K. Ståhlberg, J. Montewka & P. Kujala
Aalto University, School of Science and Technology, Department of Applied Mechanics, Espoo, Finland

ABSTRACT: Maritime traffic poses various risks in terms of human casualties, environmental pollution or loss of property. In particular, tankers pose a high environmental risk as they carry very large amounts of oil or more modest amounts of possibly highly toxic chemicals. In this paper, a simplified risk assessment methodology for spills from tankers is proposed for the Gulf of Finland, for tankers involved in a ship-ship collision. The method is placed in a wider risk assessment methodology, inspired by the Formal Safety Assessment (FSA) and determines the risk as a combination of probability of occurrence and severity of the consequences. The collision probability model is based on a time-domain micro simulation of maritime traffic, for which the input is obtained through a detailed analysis of data from the Automatic Identification System (AIS). In addition, an accident causation model, coupled to the output of the traffic simulation model is proposed to evaluate the risk reduction effect of the risk control options. Further development of the model is needed, but the modular nature of the model allows for continuous improvement of the modules and the extension of the model to include more hazards or consequences, such that the effect of risk control options can be studied and recommendations made. This paper shows some preliminary results of some risk analysis blocks for tanker collisions in the Gulf of Finland.

1 INTRODUCTION

In recent years, the volume of maritime traffic has significantly increased in the Gulf of Finland, especially because of the expansion of the Russian oil exports from harbors such as Primorsk and Vysotskiy. Up to the recent economic recession, the volume of oil exported from Russia has increased every year, and it is expected to keep increasing in the future (Kuronen et al. 2008, Helcom 2010). With this increasing traffic density, inherent risks such as oil spills are of special concern due to the highly vulnerably marine ecosystem of the Gulf of Finland (Helcom 2010).

Analysis of historic shipping accidents show that worldwide, groundings, collisions and fires are the most common accident types (Soares 2001), while in the shallow, island-littered waters of the Gulf of Finland, groundings and ship-ship collisions are most frequent (Kujala et al. 2009). This justifies the concern of this paper with the risk of oil tankers involved in ship-ship collision accidents.

The main driving idea of the model presented in this paper is the societal trend towards science-based risk-informed decision making, an idea supported by organizations such as the IMO or IALA. In the mari-time field, the Formal Safety Assessment provides a framework for this aim.

2 OUTLINE OF RISK ASSESSMENT METHODOLOGY

The risk assessment methodology is rooted in the commonly accepted framework of the Formal Safety Assessment (FSA) (Kontovas and Psaraftis, 2005). The conceptual FSA-methodology is shown in Fig. 1. It starts with an identification of hazards, followed by an analysis of the risk. Thereafter, risk control options are defined, the effect of which should be evaluated using the risk analysis method.

This should be followed by a cost-benefit analysis and recommendations as to which risk control options to implement. It is therefore essential that the risk analysis methodology is able to provide a reliable evaluation of the effect of the risk reducing measures.

The system risk is defined based on the definition of Kaplan (1997) as a set of triplets:

$$\{(s_i, l_i, c_i)\}, i=1, 2, 3,... \qquad (1)$$

Here, s_i defines the context of the accident scenario, l_i the likelihood of the accident occurring in that scenario and c_i the evaluation of the consequence in the scenario.

Fig. 1. General outline of FSA methodology

It is important to indicate that l_i and c_i are dependent on the accident scenario s_i, which is to be seen as a multi-parameter set, i.e. a range of variables relevant to the evaluation of the accident probability l_i and the consequence c_i.

The risk analysis methodology is based on a system simulation of the maritime traffic in a given area. The overall flowchart, focusing on the risk of ship collision, is shown in Fig. 2. The various modules of this model, insofar these are already available, will be introduced below.

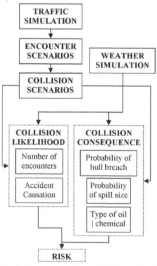

Fig. 2. General outline of FSA methodology

At present, the model is capable only to assess the risk of ship-ship collision, which is the second most important hazard in the Gulf of Finland, based on the accident statistics of Kujala et al. (2009). The

methodology can in principle be extended without too many difficulties to other accident types such as ship grounding and fires.

3 TRAFFIC SIMULATION AND COLLISION ENCOUNTER SCENARIO MODEL

The traffic simulation and collision encounter scenario detection module is one of the core units of the overall risk assessment model. The basic idea is to simulate the traffic on a micro-scale. For each vessel sailing in the area, the trajectory is simulated, while assigning a number of parameters to this vessel. These include departure time, ship type, length, loading status, cargo type and ship speed, as illustrated in Fig. 3. The simulation of all vessels in the area provides a traffic simulation and the subsequent detection of the vessels which collide, assuming that no evasive action is made, results in the definition collision encounter scenarios.

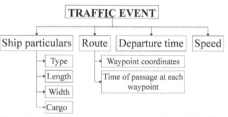

Fig. 3. Generated data for each simulated vessel (traffic event)

The input for this model is taken from data from the Automatic Identification System (AIS), augmented with statistical data from harbors concerning the traded cargo types. Details on how this simulation and collision candidate detection is performed, is given in Goerlandt and Kujala (2011).

As an illustration of the input for the simulation model, Fig. 4 shows the departure time distribution for vessels sailing from Helsinki to Tallinn. Fig. 5 shows the ship length distribution for tankers to Sköldvik and Primorsk. Fig. 6 shows the average ship speed distributions for all considered ship types. This information is used as a first estimate of the ship speed before the collision candidates are obtained. After detection of a collision candidate in a specific area, the speed is resampled from ship type specific speed distributions by location, as shown in Fig. 8. This speed is then updated in the collision encounter scenario.

Table 1 shows the harbor-specific data for cargo types of chemical tankers, for the port of Hamina. The cargos carried by the simulated vessels are sampled from this information, after a more in-depth analysis of which trade routes represent which cargo types. Fig. 7 shows the simulated traffic in the Gulf of Finland, based on the input obtained from AIS.

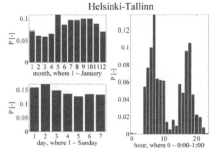

Fig. 4. Departure time distributions, traffic from Helsinki to Tallinn

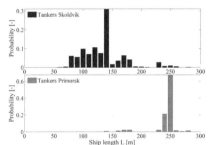

Fig. 5. Length distribution of tankers to Sköldvik and Primorsk

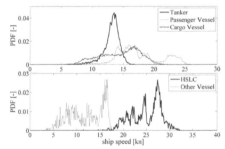

Fig. 6. Speed distributions of vessels in the Gulf of Finland

In Table 2, an example of output obtained from the collision encounter simulation model is shown. This is to be interpreted as the accident scenario context using the definition of Kaplan (1997) as presented in Section 2.

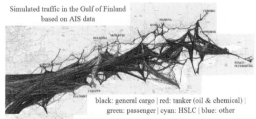

Fig. 7. Simulated traffic for one year
map: © Merenkulkulaitos lupa nro 1321 / 721 / 200 8

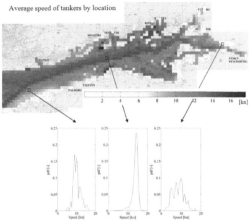

Fig. 8. Average speed of tankers in the Gulf of Finland and local speed distributions, based on AIS data of 2006-2009
map: © Merenkulkulaitos lupa nro 1321 / 721 / 200 8

Table 1. Example of data concerning harbor-specific trade volume: port of Hamina, Finland (Hänninen and Rytkonen, 2006)

IMPORT PRODUCTS

Product	Vol. [ton]	Product	Vol. [ton]
Butadiene	53926	Sulphuric acid	39492
Buthyl acrylate	12233	Styrene monomer	3380
Phenol	1038	Vinyl acetate	1457
Caustic Soda	78547	Methyl ketone	501

EXPORT PRODUCTS

Product	Vol. [ton]	Product	Vol. [ton]
Butane	741	Methyl-butyl ether	83104
Isoprene	8271	Nonylphenol	48830
Methanol	762012	Propane	2839
Styrene monomer	9602	Vinyl acetate	457
Propylene	5897		

COMMON ORIGINS	**COMMON DESTINATIONS**
St. Petersburg	Rotterdam, Antwerpen, Teesport, Hamburg, Gdynia

Table 2. Examples of encounter scenarios obtained by the model of Goerlandt and Kujala (2011)

Location [long | lat]	Time [m.h:m]	Origin	Type‡ Struck	Striking	Speed [kn]
24.60|59.82	01.05:10	Hamina	C	P	V_{loc} †
22.31|59.34	03.08:47	Sköldvik	GC	GC	V_{loc}
27.96|60.17	03.13:32	Kotka	OT	GC	V_{loc}
24.10|59.55	04.21:10	St. Petersb	GC	OT	V_{loc}
25.23|57.53	06.09:05	Vyborg	P	GC	V_{loc}
29.11|59.95	07.14:13	St. Petersb	GC	GC	V_{loc}

† V_{loc} is the local speed distribution for the relevant ship types
‡ Type: C = chemical tanker, P = passenger vessel, GC = general cargo ship, OT = oil tanker

4 COLLISION SCENARIO AND WEATHER MODEL

While the collision encounter scenario model is able to partly define the accident context, this is insufficient to accurately define either the likelihood of the accident l_i or the consequences c_i.

As a first concern, it should be noted that an encounter scenario, which depends only on the nature of the maritime traffic flows, is not equivalent to the actual collision scenario. In particular, due to possible evasive maneuvers made prior to collision, essential parameters such as vessel speed and collision angle may deviate significantly from the encounter conditions. This has an important effect on the evaluation of the consequences c_i, as can be evaluated by inspecting the collision energy models of Zhang (1999) or Tabri (2010).

Several authors have proposed models for the parameters relevant to the collision scenario, usually based on accident statistics. Some of these proposals are briefly described in Table 3. However, at present no reliable model exists linking the encounter scenario and the collision scenario. This has been investigated by Goerlandt et al. (2011) using a comparison of the hull breach probability for various collision scenario models, based on a collision energy model by Zhang (1999) and a criterion for the critical energy the ship hull can withstand before breach of the double hull. The results of the local probability of oil spill resulting from the various collision scenarios from Table 3, is shown in Fig. 9.

Table 3. Impact scenario models available in literature

Impact model by Rawson (1998)

Collision angle:	U(0,180)
$V_{striking}$:	Truncated bi-normal N(5,1) | N(10,1)
V_{struck}:	Idem as $V_{striking}$
Collision location:	U(0,180)

Impact model by NRC (2001)

Collision angle:	N(90,29)
$V_{striking}$:	W(6.5, 2.2)
V_{struck}:	E(0.584)
Collision location:	B(1.25,1.45)

Impact model by Lützen (2001)

Collision angle:	T(0, α_{enc}, 180)
$V_{striking}$:	Below .75V_{enc}: U(0, .75V_{enc})
	Above .75V_{enc}: T(.75 V_{enc}, V_{enc})
V_{struck}:	T(0, V_{enc})
Collision location:	Empirical distribution, see Lützen (2001)

† U: uniform \| N: normal \| W: weibull \| E: exponential \| B: beta \| T: triangular distribution

For a proper formulation of the accident context, a weather model, capable of predicting the factors which are needed in the evaluation of the accident likelihood and consequences, is needed as well. These factors include wind velocity, sea state and visibility. At present, this weather simulation module has not been implemented in the presented maritime accident assessment methodology.

In terms of the parameters defining the accident context, denoted s_i in the formulation of Kaplan (1997), the weather model adds certain parameters

to the values obtained from the collision scenario model, as given in Table 2. These weather-related factors affect the likelihood of the accident l_i and the effectiveness of response to oil spill.

The collision scenario model adds certain parameters such as which is the striking and struck ship and the location of the collision along the struck ship hull. In addition, this model should modify certain parameters such as the collision angle and vessel speed of striking and struck vessel, which have an important contribution to the consequence assessment, i.e. c_i in the Kaplan-nomenclature of Eq. 1.

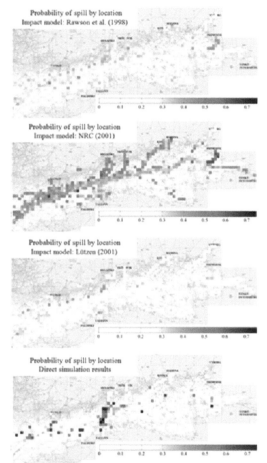

Fig. 9. Results of location-specific spill probability according to algorithm in Fig. 9 and (Eq. 4), impact models: see Table 5, map: © Merenkulkulaitos lupa nro 1321 / 721/ 200 8, taken from Goerlandt et al. (2011).

5 ACCIDENT CAUSATION MODEL

The accident causation model gives a probability of a collision accident occurring in a given context, in terms of the system risk definition by Kaplan (1999), this is the scenario specific likelihood of accident l_i.

This accident causation module from Fig. 2 is constructed using the methodology of the Bayesian Belief Network (BBN). The model is shown in Fig. 10, and is discussed in more detail in (Hänninen and Kujala 2009, Hänninen and Kujala 2011).

The model is rooted in expert opinion, accident and incident data, with the understanding that some parameters are taken directly from the output of the simulation model, in particular the traffic encounter scenario model and the weather model. For instance, the values for the nodes for encounter type, ship types and sizes, time of year, daylight condition and whether or not the encounter location is in a VTS area can be derived from the encounter scenario module as explained in Section 3. The visibility and weather conditions could be derived from the weather model as discussed in Section 4.

Table 4 gives an overview of the groups of nodes in the Bayesian Network, giving a number of examples of some nodes in these groups. The parameters which are directly taken from the traffic and weather simulation models are marked in italics.

The accident causation model is an important element in the study of the risk control options, as discussed in more detail in Section 7.

Table 4. Node groups in the Bayesian model with examples

Visual detection	Management factors
Visibility	Safety culture
Other ship size	Maintenance routines
Bridge view	Bridge resource management
Daylight	

Navigational aid detection	Human factors
Radar detection	Stress
AIS installed	Competence
AIS signal on radar screen	Situational assessment
Collision avoidance alarms	Familiarization

Support	Evasive actions / overall
VTS vigilance	*Encounter type*
Pilot vigilance	*Give way situation*
Other internal vigilance	*Time of year*
	Weather
	Ship type

Technical reliability
Steering failure
Radar functionality
AIS functionality

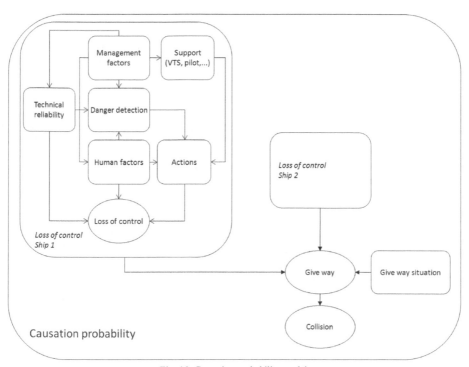

Fig. 10. Causation probability model

6 HULL BREACH PROBABILITY AND SPILL SIZE

In terms of collision consequences c_i, the focus of this paper is limited to the probability of spills from oil tankers. The environmental or socio-economic damage or implications for oil combating operations is at present not considered.

The hull breach probability can be determined based on a comparison of the available deformation energy in the collision scenario, compared to the energy which the ship structure can withstand before the inner hull is breached.

For the available deformation energy, a number of models is available. Zhang (1999) proposed a relatively simple analytical model, assuming rigid bodies and 2-dimensional ship motions. Brown (2002) proposed a simplified model taking the interaction between inner mechanics (i.e. the structural deformation) and the outer mechanics (i.e. the ship motions in a collision) into account, limited to 2-dimensional ship motions. Tabri (2010) proposed a full 6 degrees of freedom model, coupling inner and outer mechanics and taking the sloshing of liquids in a tank into account.

For the ship structural energy, methods such as finite element calculations, e.g. as proposed by Ehlers (2010) could be used. In Goerlandt et al. (2011), a simple criterion based on regression of available ship structural data is proposed.

The methodology to compare the available deformation energy and the hull structural strength is outlined in detail in Goerlandt et al. (2011). Also the work by Klanac et al. (2010) uses a variation of this approach to assess the hull breach probability.

For the oil spill size, a number of models has been proposed in the literature. Examples are a probabilistic extension of the IMO-tanker design criteria as proposed by Montewka et al. (2010). A related methodology has been proposed by Smailys and Cesnauskis (2006). A simple oil volume outflow model based on statistics of tank sizes has been proposed by Gucma and Przywarty (2007). While these models have their merits, they are very simplified and do not take the detailed information from the accident scenario s_i into account.

On the other hand, the model proposed by van de Wiel and van Dorp (2009) is capable of predicting the size of both a cargo oil spill and of a bunker oil spill using a number of variables determined in the collision accident scenarios. Such variables are the vessel sizes, speeds, collision angles and collision location along the struck ship hull. Thus, this model provides a good match with the information of the accident scenario information s_i. However, as indicated in Section 4, there exists a significant uncertainty concerning the validity of the available models for the collision scenarios.

The model is based on a combination of collision energy calculations, used to determine the damage length and width, and a limited reference ship database. Based on this information, it is assessed whether or not the hull is breached, and in the case it is, how much oil will flow out of the ship. The model can also be used for estimating the spill in case of grounding.

Since chemical tankers have a significantly different structural arrangement, the above mentioned methods can not directly be used for estimation of spill sizes of this vessel type. Also consequence evaluation based on structural damage for other vessel types is at present not available. However, the principle behind the methods proposed by Ehlers (2010) and Tabri (2010) can be used to get reliable results for these accident types.

7 OVERALL RISK ASSESSMENT: APPLICATION

The application of the risk assessment methodology is to be done by modifying the values for the risk control options in the model to evaluate the effect on the risk level. With an estimate of the cost of implementation of the risk control options and the saved cost due to the reduced risk, an informed decision can be made.

A number of risk control options are related to the accident likelihood l_i. For instance, the VTS vigilance, pilot vigilance, safety culture, navigator competence, navigational equipment and aids to navigation are taken into account in the accident causation model, as described in Section 5. Also the ship routing affects the accident likelihood, which in principle can be studied by modification of the traffic streams in the traffic simulation model, resulting in less and/or safer encounters.

Other risk control options affect the severity of the consequences in case of an accident. Examples of these are the speed limits in local sea areas and the encounter situation, which directly affect the available collision energy. Also the structural strength of the ship hull is an important factor in the severity of the consequences. The accident response effectiveness in terms of number, location and equipment of the available oil response or search and rescue fleet, can also be studied based on the risk maps produced in the risk analysis step.

It should be noted in this context that estimating the accident costs is a difficult task in itself due to the highly complex nature of the studied system. For instance, for an oil spill due to collision, apart from the spill size, the ecological and socio-economic value of the environment in which the spill may occur, should be considered.

It may therefore be more feasible to study the relative risk reduction of the measures as such, and comparing these to the costs of the risk control options. This will also lead to a decent risk-informed decision, if a certain expertise is available to interpret the results of the risk assessment.

8 CONCLUSION AND FUTURE WORK

It should be clear that the evaluation of the risks related to maritime traffic in the framework of a Formal Safety Assessment is a very wide and laborious task, not in the least because of the multidisciplinary nature of the studied system. Such fields as logistics, maritime engineering, systems analysis, operations research and environmental modeling should be combined in an overall FSA-framework.

The maritime system simulation methodology starts from the premise that the likelihood and consequence of each relevant accident type can be calculated based on situational information, as suggested by Kaplan (1997). The aim of determining each of the modules building up the model for maritime system risk in a scientifically sound manner is to be seen as an attempt to rationalize the decision making process in risk related matters.

It is clear that even though the scope of the current model is rather limited (only the probability of collision of ships in open waters and the consequences in terms of oil spill size are included as yet), and even within these models certain improvements could be made (e.g. the collision scenario model linking encounter scenario to actual impact conditions), the modular nature of the model allows for gradual improvement and extension of the models to include additional hazards, risk analysis blocks or risk control options.

Consequently, the remaining work is still very significant before any proper conclusions can be made. Firstly, other hazards (ship grounding, fire) should be included. Secondly, a weather model should be coupled to the accident scenario generation. Thirdly, for ship collisions, the consequences for other ship types (chemical tankers, passenger vessels), should be determined in terms of economic loss due to structural damage or loss of human life. There is also significant work to be done in the understanding of accident causation, and for various accident types, there is a lack of consequence models.

ACKNOWLEDGEMENTS

The authors appreciate the financial contribution of the European Union and the city of Kotka. This research is carried out within the EfficienSea project and in association with the Kotka Maritime Research Centre.

REFERENCES

Brown, A.J., 2002. Collision scenarios and probabilistic collision damage. *Marine Structures, 15(4-5):335-364.*

Ehlers, S. 2010. Material relation to assess the crashworthiness of ship structures. *Doctoral Dissertation, Aalto University, School of Science and Technology.*

Goerlandt, F., Kujala, P. 2011. Traffic simulation based ship collision probability modeling. *Reliability Engineering and System Safety* doi:10.1016/j.ress.2010.09.003.

Goerlandt, F., Ståhlberg, K., Kujala, P. 2011. Comparative study of input models for collision risk evaluation. *Ocean Engineering – manuscript under review.*

Gucma, L., Przywarty, M. 2007. The model of oil spills due to ships collisions in Southern Baltic area. *TRANSNAV 2007, Gdynia, Poland.*

Hänninen, M. & Kujala, P. 2009. The Effects of Causation Probability on the Ship Collision Statistics in the Gulf of Finland. *In Weintrit, A. (ed.) Marine Navigation and Safety of Sea Transporation, 267-272.*

Hänninen, M. & Kujala, P. 2011. Influences of variables on ship collision probability in a Bayesian belief network model. *In prep.*

HELCOM. 2010. Maritime Activities in the Baltic Sea – An integrated thematic assessment on maritime activities and response to pollution at sea in the Baltic Sea Region. *Balt. Sea Environ. Proc. No. 123.*

Kaplan, S., 1997. The Words of Risk Analysis. *Risk Analysis, 17(4), 407-417.*

Klanac, A., Duletic, T., Erceg, S., Ehlers, S., Goerlandt, F., Frank, D. 2010. Environmental risk of collision for enclosed seas: Gulf of Finland, Adriatic and implications for tanker design. *5th International Conference on Collision and Grounding of Ships, Espoo.*

Kontovas, C., Psarafitis, H. 2005. Formal Safety Assessment – Critical review and future role. *Msc. Thesis, National University of Athens.*

Kujala, P., Hänninen, M., Arola, T., Ylitalo, J. 2009. Analysis of the marine traffic safety in the Gulf of Finland. *Reliability Engineering and System Safety 94(8):1349-1357.*

Lützen, M., 2001. Ship collision damage. *PhD thesis, Technical University of Denmark.*

Montewka, J., Ståhlberg, K., Seppala, T., Kujala, P. 2010. Elements of risk analysis for collision of oil tankers. In G. Soares (Ed.), *ESREL 2010,* London. Taylor and Francis.

National Research Council (NRC), 2001. Environmental Performance of Tanker Designs in Collision and Grounding. *Special Report 259, The National Academies Press.*

Smailys, V., Cesnauskis, M. 2006. Estimation of expected cargo outflow from tanker involved in casualty. *Transport Vilnius 21(4):293-300.*

Tabri, K. 2010. Dynamics of ship collisions. *Doctoral Dissertation, Aalto University, School of Science and Technology.*

van de Wiel, G., van Dorp, J.R. 2009. An oil outflow model for tanker collisions and groundings. *Annals of Operations Research:* doi: 10.1007/s10479-009-0674-5.

Zhang, S., 1999. The mechanics of ship collisions. *PhD thesis, Technical University of Denmark.*

26. Estimating the Number of Tanker Collisions in the Gulf of Finland in 2015

M. Hänninen, P. Kujala & J. Ylitalo[*]
Aalto University School of Engineering, Espoo, Finland

J. Kuronen
University of Turku, Centre for Maritime Studies, Kotka, Finland

ABSTRACT: The paper presents a model for estimating the number of ship-ship collisions for future traffic scenarios. The modeling is based on an approach where the number of collisions in an area is estimated as a product of the number collision candidates, i.e. the number of collisions of two ships, if no evasive maneuvers were made, and a causation probability describing the probability of making no evasive maneuvers. However, the number of collisions is presented as a combination of binomially distributed random variables. The model is applied for the assessment of tanker collision frequency in the Gulf of Finland in 2015. 2015 traffic is modeled as three alternative scenarios each having a certain probability of occurrence. The number of collisions can be presented either for each scenario, or as an estimate including the uncertainty in future marine traffic development by taking into account all scenarios and their occurrence probabilities.

1 INTRODUCTION

The Gulf of Finland is a highly trafficked area in the Baltic Sea. Moreover, its traffic is expected to grow in the future (e.g. Hassler 2010; Kuronen et al. 2009; Ministry of Transport and Communications Finland 2009). Growing traffic increases the risk of ship groundings and collisions, which are the two most common types of maritime traffic accidents in the Gulf of Finland (Kujala et al. 2009). Especially the increasing oil tanker traffic and the possibility of a major oil accident raise concern in the coastal states.

One of the most common approaches to estimating the number of ship collisions was introduced by Fujii et al. (1971) and Macduff (1974). In this approach the number of collisions N is calculated as a product of the number of geometrical collision candidates N_G and a causation probability P_C:

$$N = N_G \times P_C \qquad (1)$$

N_G describes the number of collisions of two ships, if they do not perform evasive maneuvers. It is based on traffic properties of the area, such as the routes of the ships, ship particulars (length, width) and velocities. A few models for assessing N_G are existing, such as Pedersen's model (1995) and the MDTC model (Montewka et al. 2010b), or it can be estimated with time-domain micro simulation (Goerlandt et al. 2011). P_C describes the probability of collision candidate ships making no evasive maneuvers. It is determined by various factors affecting human and/or technical failure. The causation probability

has been estimated based on the difference between accident frequencies according to accident statistics and the estimated number of collision candidates (Fujii 1971; Macduff 1974), or by applying risk analysis tools such as fault tree analysis (Pedersen 1995, Rosqvist et al. 2002). Bayesian belief networks have been suggested to be utilized in in Step 3 of the Formal Safety Assessment, definition of risk control measures (IMO 2006), and more recently they have been applied in causation probability estimation (e.g. Friis-Hansen and Simonsen 2002, Det Norske Veritas 2003, Rambøll 2006, Hänninen & Ylitalo 2010).

Recently, in several studies the collision probabilities in the Gulf of Finland have been examined with the approach of Equation 1. Montewka et al. (2010a) estimated the number of geometrical collision candidates for a crossing between Helsinki and Tallinn using). The authors of this paper (Hänninen & Kujala 2009) estimated the number of collisions for the same crossing area. Later, the authors (Hänninen & Ylitalo 2010) estimated the collision frequency for the whole Gulf of Finland.

The studies mentioned above had not assessed the future risks in the Gulf of Finland. Further, all of the mentioned studies had estimated the number of collision candidates or collisions as point estimates. In this study, a probability distribution for the number of tanker collisions in the Gulf of Finland is presented. The number of tanker collisions is estimated for 2015 traffic by utilizing AIS data and three alternative maritime transportation growth scenarios. The

---

[*] affiliation at the time of contribution

study also considers the effects of uncertainty in the occurrence of the 2015 scenarios.

2 STUDIED AREA

The main waterways in the Gulf of Finland were included in the analysis. The waterways were defined based on a traffic image from AIS data. However, areas within the vicinity of ports were excluded. Additionally, tanker collisions within four smaller areas of the gulf were studied separately. The studied waterways and the "hot spot" areas are presented in Figure 1. The considered "hot spot" areas were: C1: the crossing of Helsinki-Tallinn traffic and the main route of the Gulf of Finland; C2: the merging of Sköldvik and the main route traffic,; C3: the merging of traffic of Primorsk and St. Petersburg and the waterway to St. Petersburg; and C4: the westernmost part of the Gulf. C1 was chosen due to high traffic within the area. C2 was considered as a possible collision area for two tankers on oil load. C3 included a rather narrow waterway to St. Petersburg and a merging of two waterways near shoals. C4 was an example of a larger area with many crossing and merging waterways.

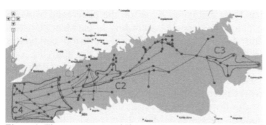

Figure 1. The waterways and "hot spot" areas whose number of tanker collisions were estimated.

3 METHODS

3.1 *2015 traffic estimation*

The estimates for the numbers of ships in the studied waterways in 2015 were based on 2008 traffic and traffic multipliers extracted from growth scenarios for the ports in the Gulf of Finland. AIS data from the area was utilized in determining the traffic in 2008. It should be noted that the winter of 2008 was exceptionally mild, and no ice breaking assistance was needed in the Gulf of Finland. Thus, the results are describing open water season only.

For the 2015 traffic, three alternative scenarios were considered: "slow growth", "average growth" and "strong growth" (Kuronen et al. 2009). The expected value of the total tonnes for the maritime transportation in the Gulf of Finland in "slow growth" was 322 million tonnes. For the "average growth" scenario, the number was 432 M tonnes,

and for the "strong growth", 507 M tonnes. The growth factors for each port in three scenarios were defined as follows. First, the total amount of oil and other cargoes were defined to each scenario (see Kuronen et al. 2009). Second, the amount of oil and other cargoes were distributed to the ports according to the shares of the cargo amounts in the ports in 2007. Third, these port distributions were modified on the basis of the expertise of the research group, taking into consideration e.g. Ust-Luga port building project, other expected changes in the traffic patterns and the basic assumptions concerning the development of the traffic in each scenario. The growth factors are presented in Table 1.

One should note that the growth scenarios were based on the transport in 2007, whereas the AIS traffic multiplied with the traffic multipliers was from 2008. According to AIS data, the number of ship movements at the entrance to the gulf had decreased by 6.0 % from 2007 to 2008. However, the decrease was not constant in the whole Gulf of Finland: for example, the change was smaller on the main route on the eastern side of Gogland, and the number of passenger vessels even grew on that particular waterway. Overall, the magnitude of the change was only approximately 5 %, and considering other sources of uncertainty related to future traffic prediction, it was decided to define the multipliers based on the growth scenarios from 2007 traffic.

The traffic in 2008 was multiplied with waterway-specific multipliers to obtain estimates for the traffic in the waterways in 2015. Based on the port-specific growth factors, a cargo volume multiplier and an oil volume multiplier were calculated for each segment of each waterway included in the study. For waterways leading to ports, the multipliers were equal to the port's growth factor. For other waterways, multipliers were deduced as a combination of the multipliers of merging waterways in relation to the traffic volumes. The traffic distributions across each waterway were assumed to remain unchanged. In addition, the ship size distribution was assumed not to change. No changes were made to the numbers of passenger ships, high speed crafts, and other ships, as no similar estimates on the change of their volume were available. Percentages of traffic continuing to separate waterways at waypoints were also adjusted to the changed traffic volumes on the waterways.

It should be noted that no oil was transported from Ust-Luga in 2007, so the number and size of tankers navigating there were obtained by assuming the size distribution of tankers being similar to that of the tankers navigating to St. Petersburg in 2008. The estimated number of tankers was added entirely to the eastern waterway to Ust-Luga since all tankers had used it in 2008.

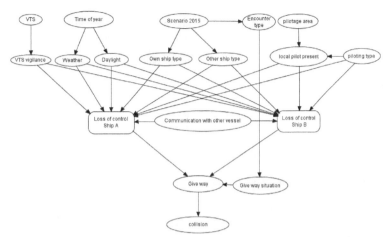

Figure 2. The structure of a Bayesian network model for the causation probability estimation.

Table 1. Cargo and oil volume growth factors from 2007 to 2015 for the scenarios "Slow" (C_{sl} and O_{sl} for cargo and oil, respectively), "Average" (C_{av} and O_{av})and "Strong" (C_{st} and O_{st}).

Port	C_{sl}	O_{sl}	C_{av}	O_{av}	C_{st}	O_{st}
Helsinki	1.01	1.03	1.31	1.05	1.34	1.35
Sköldvik	1.16	1.01	1.10	1.10	0.00	1.18
Kotka	1.01	1.00	1.32	0.97	1.35	1.37
Hamina	1.02	1.00	1.38	1.26	1.40	1.45
Hanko	1.47	1.00	1.47	1.00	1.53	1.00
Vysotsk	1.16	1.06	1.26	1.12	1.40	1.24
Primorsk	-	1.35	-	1.62	-	1.62
St. Petersburg	1.12	1.00	1.37	1.37	1.43	1.43
Ust-Luga	6.41	-	8.54	-	11.29	-
Sillamäe	1.00	0.00	1.06	1.06	1.61	1.61
Tallinn	1.33	0.01	1.84	0.63	2.39	1.19

3.2 Collision probability modeling

The number of collisions was calculated separately for the encounters of two oil tankers, and the encounters of an oil tanker and another type of ship. The expected value for the number of tanker collision candidates for the 2015 traffic scenarios were calculated with IWRAP software. IWRAP is recommended for the evaluation of collision probabilities by International Association of Marine Aids to Navigation and Lighthouse Authorities (IALA 2009). The calculations were performed to all three traffic scenarios and to all considered areas in a similar manner as the authors had done for the 2008 traffic in the whole Gulf in (Hänninen & Ylitalo 2010), which gives a more detailed description of the method.

The causation probability was modeled and estimated with a Bayesian belief network. Bayesian networks are directed acyclic graphs which consist of nodes representing discrete random variables and arcs representing the dependencies between the variables (e.g. Jensen & Nielsen 2007). Each variable consists of a finite set of mutually exclusive states. For each variable A with parent nodes $B_1,..., B_n,$

there exists a conditional probability table $P(A \mid B_1, ..., B_n)$. If variable A has no parents, it is linked to unconditional probability $P(A)$. The model applied in this study was partly based on a collision model network in the Formal Safety Assessment of large passenger ships (Det Norske Veritas 2003) and a grounding model in the FSA of ECDIS chart system (Det Norske Veritas 2006). Additionally, expert knowledge and data from the Gulf of Finland ship traffic and environmental conditions were used in constructing the model. More detailed description of the model variables and the probability parameters can be found in (Hänninen & Kujala, in prep.), where the authors have described a more detailed causation probability model with many similarities to the model applied in this study. A variable "Scenario 2015" with states "slow", "average" and "strong" describing the degrees of belief of the traffic scenarios' occurrence was added to the model. In the causation probability model, the traffic scenario was directly influencing only the ship type and encounter type distributions.

The model was constructed as an object-oriented Bayesian network (OOBN). OOBN enables the use of sub classing in Bayesian network models (Jensen & Nielsen 2007). If a Bayesian network model contains a repetitive substructure, a separate sub model or a class could be constructed from this network substructure. In an OOBN, several instances of this class can then be inserted into the main model as instance nodes. This enables data abstraction, i.e., hiding the more detailed variables and their dependencies inside a class whose input and output are only visible in the main model. In this study, a class was constructed from the set of variables and arcs describing a ship losing control, and two instances of this "Loss of control" sub model were then created within the main model, describing the loss of control for each of the two meeting ships. "Own ship type" distribution was given as an input to the variable "Own ship type" in the instance of "Loss of control" sub model for the "ship A", and as an input to "Other ship type" for the "ship B". The main model is presented in Figure 2, and Figure 3 describes the network structure of the "Loss of control" sub model.

After calculating the expected values for the number of collision candidates and the causation probability as is described above, the number of collisions N within a year was modeled with a binomial distribution. Binomial distribution is a discrete probability distribution for the number of successes in certain number of independent yes/no experiments, when success in one experiment occurs with a certain probability. For the number of collisions distribution, the number of experiments was the number of collision candidates, and the probability of one success was the causation probability. Thus the probability of having exactly n collisions was

$$\Pr(N = n) = \frac{N_G!}{n!(N_G - n)!} P_C^n (1 - P_C)^{N_G - n}, \quad (2)$$

where N_G is the number of geometrical collision candidates and P_C is the causation probability. Finally, the distributions of the number of "tanker-tanker" collisions and the number of "tanker-not tanker" collisions were combined in order to acquire the number of collisions where at least one tanker was involved.

4 RESULTS

The expected values of the number of collisions involving at least one tanker in the whole Gulf of Fin-

land and in the "hot spots" for the three 2015 traffic scenarios are presented in Table 2. With the "average" scenario, the expected yearly number of tanker collisions in the Gulf of Finland was estimated to be 0.17, which equals one tanker collision within approximately six years. If the "hot spots" are considered, the largest expected collision probability was in the area C3, including the merging waterways of Primorsk and St. Petersburg. For the "average" scenario, 0.044 tanker collisions were estimated to occur there within a year, which equals a collision in every 23 years.

Table 2. The expected values of the number of collisions / year involving at least one tanker for the 2015 traffic scenarios "Slow", "Average" and "Strong".

Area	Slow	Average	Strong
GoF	0.127	0.173	0.183
C1	0.010	0.012	0.139
C2	0.016	0.021	0.023
C3	0.033	0.044	0.044
C4	0.011	0.014	0.016

Table 3. The mean time (years) between collisions involving at least one tanker for scenario combinations with various weightings ("Slow-Average-Strong").

Area	0.33-0.33-0.33	0.35-0.5-0.15	0.15-0.5-0.35
GoF	6.2	6.3	5.9
C1	83.9	86.2	80.2
C2	50.2	51.6	48.0
C3	24.8	24.9	23.6
C4	74.7	76.8	71.7

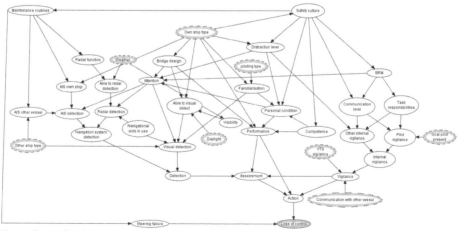

Figure 3. "Loss of control" sub model.

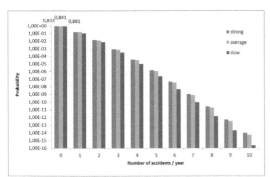

Figure 4. Probability distribution of the number of tanker collisions in the whole Gulf of Finland in 2015 for the three traffic scenarios "slow", "average" and "strong". The probability values of having zero collisions is presented above the corresponding bars.

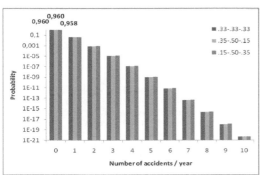

Figure 5. Probability distribution of the number of tanker collisions in the area C3 in 2015 for the scenario combinations with various weightings of the scenarios.

Figure 4 presents the probability distribution of the number of tanker collisions within a year in the whole Gulf given for all traffic scenarios. The probability of having zero tanker collisions within a year was between 0.83 and 0.89, depending on the scenario. The number of tanker collisions can also be examined while taking the uncertainty of the occurrence of the traffic scenario into account. This was done by assigning a weight to each of the scenarios. The weight was describing the degree of belief in the occurrence of the traffic scenario in question, assuming that the "true" scenario is amongst the three alternatives, i.e., the weights sum up to 1.0. Table 3 presents the mean time between tanker collisions in the areas with various weightings of the scenarios: all scenarios equally likely to occur, and two other alternatives, where "average growth" was assigned 0.5 weighting, and 0.15/0.35 weights were assigned to the other scenarios. These weightings were identical to the ones experts had assigned to the scenarios in (Kuronen et al. 2008). Figure 5 presents the probability distributions of the number of tanker collisions for the specific weightings. As can be seen

from Table 3 and Figure 5, the differences in weighting had a minor effect on the outcome.

5 CONCLUSIONS

In this study, a probability distribution of the number of collisions in the future given uncertainty in maritime traffic development was presented. The model was applied to the Gulf of Finland maritime traffic growth scenarios. The number of "tanker-tanker" collisions and "tanker-not tanker" collisions were modeled separately using binomial distributions and then combined. According to the results, a collision involving at least one tanker would occur once in approximately every six years. This might seem a rather high number, especially since tanker collisions in the Gulf of Finland within open water season have been quite rare (Hänninen & Ylitalo 2010). Nevertheless, it should be noted that the "average growth" scenario, for example, is estimating a 60 % increase in transportation tonnes compared to the traffic in 2007 (Kuronen et al 2009). Further, the increase is mainly due to increase in oil transport. Therefore, there should also be an increase in the probability of tanker collisions.

The "hot spot" area with the largest estimated number of tanker collisions would be the merging area of St. Petersburg and Vysotsk traffic. This seems realistic, since according to the accident statistics of the Gulf of Finland (Hänninen & Ylitalo 2010), all non-ice related collisions had occurred in the eastern part of the Gulf.

The expected transportation tonnes in "strong growth" scenario was approximately 57 % larger than in "slow growth". Consequently, the expected value for number of collisions in the "strong" scenario is 44 % larger than in the "slow growth" scenario. In contrast, when comparing the results of 15-50-35 and 35-50-15 degree of belief weightings of the traffic scenarios, the difference is not as clear. If a weight of 0.15 was assigned to the "slow growth" scenario and 0.35 to the "strong", the expected value for number of collisions is only 5 % larger than if the weights were assigned the other way round. This can be explained by the relatively large weight given to the "average" scenario (50 %) in both cases.

The modeling of the 2015 traffic included many simplifications: the only difference between the present maritime traffic and the one in 2015 was assumed to be the numbers of oil tankers and other cargo vessels navigating in the waterways. The increase of the number of passenger ships, other ships, high speed crafts, and chemical and gas tankers navigating in the Gulf of Finland was not considered. Moreover, the change in tanker and cargo vessel numbers was estimated based on the assumption of no change in ship size. Also, the locations of the waterways were assumed to remain unchanged from the 2008 situation, and the impacts of winter on collision probability were excluded from the analysis. The changes in variables affecting the causation probability, such as the rules, regulations, safety culture and the competence of the mariners, or in technical equipment and the ships themselves, were not considered in this study and should be taken into account when building a more sophisticated model for assessing the collision risks in the future.

In Kuronen et al. (2009), each of the three traffic scenarios had been presented as probability distributions. In order to include the uncertainty in the scenarios themselves, instead of using only the expected values, the traffic multipliers could also be expressed in a distribution form. Further, considering the large number of variables with complicated interrelations behind accident causation, the quality of AIS data utilized in the traffic image composition, and selection of the models to be applied for the collision candidate and causation probability estimation, one should also address the uncertainty in the number of collision candidates and causation probability as well.

The approach presented in the paper could be utilized in a wider risk analysis and decision-making context. This work has already been started, as in Lehikoinen et al. (in prep.) the presented model was utilized as a part of a probabilistic decision analysis model of oil transportation risks, whose purpose is to aid the decision makers in choosing the best risk control options when considering the environmental consequences of oil accidents.

ACKNOWLEDGEMENTS

The study was conducted as a part of SAFGOF and CAFE projects, financed by the European Union - European Regional Development Fund - Regional Councils of Kymenlaakso and Päijät-Häme, the City of Kotka, Kotka-Hamina regional development company Cursor Ltd., Kotka Maritime Research Association Merikotka and the following members of the Kotka Maritime Research Centre Corporate Group: Port of Hamina, Port of Kotka and Arctia Shipping Oy (formerly Finstaship). The authors wish to express their gratitude to the funders.

REFERENCES

[DNV] Det Norske Veritas. 2003. Formal Safety Assessment – Large Passenger Ships, ANNEX II.

[DNV] Det Norske Veritas. 2006. Formal Safety Assessment of Electronic Chart Display and Information System (EC-DIS). Technical Report No. 2005-1565, rev. 01.

Friis-Hansen, P & Simonsen, B.C. 2002. GRACAT: software for grounding and collision risk analysis. *Marine Structures* 15(4): 383-401.

Fujii, Y. & Shiobara, R. 1971. The Analysis of Traffic Accidents. *Journal of Navigation* 24 (4): 534-543.

Goerlandt, F. & Kujala, P. 2011. Traffic simulation based ship collision probability modeling. Reliability Engineering and System Safety 96(1): 91-107.

Hassler, B. 2010. Global regimes, regional adaptation; environmental safety in Baltic Sea oil transportation. *Maritime Policy & Management* Vol. 37, No. 5, 489-503.

Hänninen, M. & Kujala, P. 2009. The Effects of Causation Probability on the Ship Collision Statistics in the Gulf of Finland. In Weintrit, A. (ed.)

Hänninen, M. & Ylitalo, J. 2010. Estimating ship-ship collision probability in the Gulf of Finland. *5th International Conference on Collision and Grounding of Ships ICCGS 2010*, 14.-16. June 2010, Espoo, Finland. TKK-AM-16: 250-255.

Hänninen, M. & Kujala, P. in prep. Influences of variables on ship collision probability in a Bayesian belief network model.

[IALA]. International Association of Marine Aids to Navigation and Lighthouse Authorities. 2009. IALA Recommendation O-134 on the IALA Risk Management Tool for Ports and Restricted Waterways, Edition 2.

[IMO] International Maritime Organization. 2006. Formal Safety Assessment. Consideration on Utilization of Bayesian Network at Step 3 of FSA, Maritime Safety Committee 81st Session, MSC 81/18/1.

Jensen, F.V. & Nielsen, T.D. 2007. *Bayesian Networks and Decision Graphs*. New York: Springer.

Kujala, P., Hänninen, M., Arola, T. & Ylitalo J. 2009. Analysis of the marine traffic safety in the Gulf of Finland. *Reliability Engineering and System Safety* 94(8): 1349-1357.

Kuronen, J., Helminen, R., Lehikoinen, A. & Tapaninen, U. 2008. Maritime Transportation in the Gulf of Finland in 2007 and in 2015. *Publications from the Centre for Maritime Studies*, A 45. University of Turku.

Kuronen, J., Lehikoinen, A. & Tapaninen, U. 2009. Maritime transportation in the Gulf of Finland in 2007 and three alternative scenarios for 2015. *International Association of Maritime Economists Conference IAME 2009 Conference*, 24-26 June 2009 Copenhagen Denmark.

Lehikoinen, A., Luoma, E., Hänninen, M. & Kuronen J. in prep. Bayesian influence diagram as integrative modeling tools in cross-disciplinary and multi-risk decision analysis of maritime oil transportation.

Macduff, T. 1974. The probability of vessel collisions. *Ocean Industry* 9(9): 144–148.

Ministry of Transport and Communications Finland 2010. Baltic Sea Maritime Safety Programme. Publications of the Ministry of Transport and Communications 18/2009.

Montewka, J., Krata, P. & Kujala, P. 2010a. Elements of risk analysis for collision and grounding of oil tankers in the selected areas of the Gulf of Finland. *5th International Conference on Collision and Grounding of Ships ICCGS 2010*, 14.-16. June 2010, Espoo, Finland. TKK-AM-16: 159-168.

Montewka, J., Hinz, T., Kujala, P. & Matusiak, J. 2010b. Probability modeling of vessel collisions. *Reliability Engineering and System Safety* 95.:573-589.

Pedersen, P.T. 1995. Collision and Grounding Mechanics. *Proceedings of WEMT'95*. Copenhagen: The Danish Society of Naval Architects and Marine Engineers. 125-127.

Rambøll. 2006. Navigational safety in the Sound between Denmark and Sweden (Øresund); Risk and cost-benefit analysis. Rambøll Danmark A/S.

Rosqvist, T., Nyman, T., Sonninen, S. & Tuominen, R. 2002. The implementation of the VTMIS system for the Gulf of Finland - a FSA study. Proceedings of the RINA International Conference Formal Safety Assessment; 2002 Sep 18-19; London,UK. 151-164.

Oil Spill Response

27. The Method of Optimal Allocation of Oil Spill Response in the Region of Baltic Sea

L. Gucma, W. Juszkiewicz & K. Łazuga
Maritime University of Szczecin, Poland

ABSTRACT: Massive catastrophic oil spills such as the Torrey Canyon (1976), Amoco Cadiz (1978), Exxon Valdez (1989) focus public attention on the damage caused by such accidents. Such massive spills are rare, but there are more small and moderate oil spills caused by vessel collisions, pipeline ruptures, and operational discharges from tankers. In case of oil spill response action must be taken to control pollution at the sea and to prevent coastline pollution. Such actions involve the dispatching of cleanup equipment. Main problem of the oil spill cleanup action seems to be question of the evaluation of the tradeoffs between cleanup and damage costs. This paper lays out the method of optimal allocation of oil spill response resources and method of costs optimize.

1 PROBLEMATIC OF OIL SPILLS AT SEA

1.1 *Traffic at Baltic Sea*

The Baltic Sea is one of the most heavily trafficked seas in the world. Ships traffic account for 15% of the world's cargo transportation. Both the number and the size of ships have grown in recent years, especially in respect to oil tankers, and this trend is expected to continue.

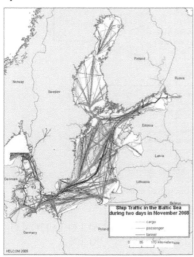

Figure 1. Cargo, tanker and passenger ship traffic on the Baltic Sea during two days in November 2008 (*HELCOM, Overview Of The Shipping Traffic In The Baltic Sea, April 2009*).

The main environmental effects of shipping and other activities at sea include air pollution, illegal deliberate and accidental discharges of oil, hazardous substances and other wastes, and the unintentional introduction of invasive alien organisms via ships' ballast water or hulls.

According to the HELCOM AIS, there are about 2,000 ships in the Baltic marine area at any given moment, and each month around 3,500–5,000 ships ply the waters of the Baltic (*HELCOM, Overview Of The Shipping Traffic In The Baltic Sea, April 2009*).

1.2 *Overview of oil spills at the Baltic Sea*

The transportation of oil and other potentially hazardous cargoes is growing steeply and steadily. More than 4,400 tankers loaded with oil left or entered the Baltic Sea in 2007 and in both 2007 and 2008 approximately 170 million tonnes of oil were shipped on the Baltic Sea. Both the number and size of the ships (especially oil tankers) have been growing during last years and now ships carrying up to 150 thousand tons of oil can be seen in the Baltic. By 2015, a 40% increase is expected in the amounts of oil being shipped on the Baltic and the number of large tankers is expected to grow, with more tankers carrying 100,000-150,000 tonnes of oil (*HELCOM, Overview Of The Shipping Traffic In The Baltic Sea, April 2009*).

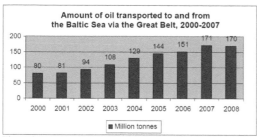

Figure 2. Amount of oil transported to and from the Baltic Sea via the Great Belt during 2000-2008 (*SHIPPOS 2000-2007 and Danish reporting system, 2008*)

2 METHOD OF OPTIMAL ALLOCATION OF OIL SPILL RESPONSE RESOURCES

2.1 *Response resources allocation at Baltic Sea*

When an oil spill occurs it is necessary to respond with sufficient cleanup equipment within the shortest possible time in order to protect marine environment and minimize cleanup and damage costs. At Baltic Sea region every country is equipped with their own response resources. Picture below (Fig.3) shows location of those equipment.

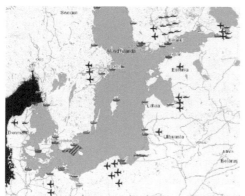

Figure 3. Response resources at Baltic Sea *(HELCOM database)*

2.2 *Model of optimal allocation*

Model apply the statistical data consist of the frequency, volume, type, location, weather conditions, and sea-state of an oil spill event. Statistical analysis is performed on historical data to determine the expected volume, type and weather and sea conditions for Baltic Sea region. The analysis is performed to determine certain input parameters such as the number and type of equipment required to respond to a given spill and the expected travel times for transporting the equipment from a facility site to spill site. The travel time for response equipment depends on the distance between facility site and the spill site, the type of equipment, and on the weather and sea conditions. After obtain the required data they are used to simulate an oil spill on PISCES II simulator.

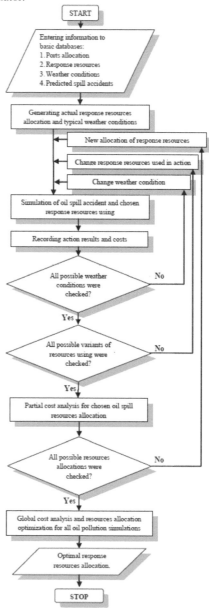

Figure 4. Model of optimal allocation.

3 COST OPTIMIZATION OF ALLOCATION OF ANTI POLLUTION RESOURCES

3.1 *Optimization model*

It is assumed that the fixed costs of opening facility site and the operating costs of the equipment are known. Costs include the costs of acquiring the equipment to be located at the site (these costs depend on the location of the site and vary with the geographical location), cost of maintaining the equipment, the transportation and cleanup costs. The following goal function is applied:

$$K_A + K_S \rightarrow min$$

where:

K_A-cost of cleanup operation,
K_S-cost of environmental pollution,

with restrictions:

O_1- allocation of resources are only in specific locations (eq. ports),
O_2- the number of available rescue units and the cost of their maintenance in standby for action does not exceed the state budget,
O3- disposal of recovered oil and oily waste should only be considered after all possibilities of processing it for use as a fuel or raw materials have been exhausted.

4 MODEL APPLICATION - CASE STUDY

4.1 *Simulation model*

PISCES II is an incident response simulator designed for preparing and conducting command centre exercises and area drills. The application is developed to support exercises focusing on oil spill response. The PISCES II provides the exercise participants with interactive information environment based on the mathematical modelling of an oil spill interacting with surroundings and combat facilities. The system also includes information-collecting facilities for the assessment of the participants' performance.

The PISCES II spill model simulates processes in an oil spill on the water surface: transport by currents and wind, spreading, evaporation, dispersion, emulsification, viscosity variation, burning, and interaction with booms, skimmers, and the coastline (stranding or beaching). The following factors are taken into consideration in the math model:
– Environmental parameters: coastline, field of currents, weather, wave height and water density;
– Physical properties of spilled oil: specific gravity, surface tension, viscosity, distillation curve and emulsification characteristics;
– Properties of spill sources;
– Human response actions: booming, on-water recovery, application of chemical dispersants.

4.2 *Simulation input data*

The simulation scenarios have been build on with application of two potential oil spill points at Baltic Sea: the first in Gdańsk Bay and the other in the vicinity of Bornholm. Those points have been chosen from stochastic oil spill risk model presented in *Gucma L., Przywarty M.: "The Model of Oil Spills Due to Ships Collisions in Southern Baltic Area"*. The *"National Plan of Fighting Threats and Environmental Pollutions at Sea"* have been also considered.

Oil spill accident can occur in arbitrary moment. Scenarios were simulated for risk of oil spill impact evaluation. Meteorological conditions represent average Baltic Sea conditions. On the base of wind and current data probable situations were formed. The data show hypothetical pollution zones for no ice conditions.

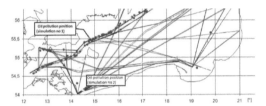

Figure 5. Risk points at the Baltic Sea *(Gucma, Przywarty TRANS'NAV 2007)*.

4.3 *Simulation no 1*

First stage of simulations was held in the vicinity of Bornholm and average spring weather conditions were used. In the accident point 6 000 ton of light oil reached the leak. Accurate data concerning spilt oil and weather conditions are described in the table 1.

Table 1. Simulation 1- weather conditions

Accident coordinates	φ=55°29,435'N
	λ=014°48,462'E
Current direction	90°
Current speed	0,25 kts
Wind direction	270°
Wind speed	8 kts
Air temperature	20 °C
Water temperature	8 °C
Sea state	1 m
Water density	1,006
Pressure	1012 hPa
Cloudiness	5
Amount of oil	6000 t
Ratio	6000 t/h
Type of oil	IFO 180

Figure below shows oil slick movement after oil spill at Baltic Sea (Bornholm).

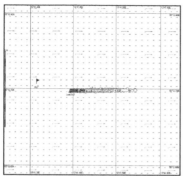

Figure 6. Oil slick after 16 min.

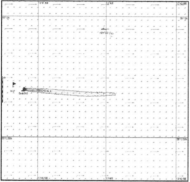

Figure 7. Response vessel under way.

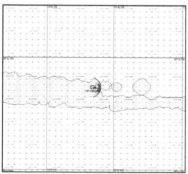

Figure 8. Movement of boom formation.

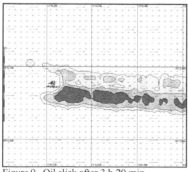

Figure 9. Oil slick after 3 h 20 min.

4.4 *Simulation no 2*

This scenario was held in Pomorska Bay and average summer weather conditions (June) were simulated. In the set point 5000 tons of oil reached the leak. Oil and weather conditions are described in the table 2.

Table 2: Simulation no 2- weather conditions.

Accident coordinates	φ=54°18,6'N
	λ=014°15,6'E
Environmental conditions	
Current direction	90°
Current speed	0,25 kts
Wind direction	100°
Wind speed	3 kts
Air temperature	20 °C
Water temperature	15 °C
Sea state	0 m
Water density	1,006
Pressure	1012 hPa
Cloudiness	5
Amount of oil	20 000 t
Ratio	5000 t/h
Type of oil	IFO 180

Boom Formation:
- Boom: RO-BOOM 1500
- Left vessel: m/s Zefir
- Right vessel: m/s Czesław
- Skimmer: Seaskimmer 50
- Middle vessel: m/s Kapitan Poinc

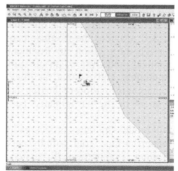

Figure 10. Simulation start.

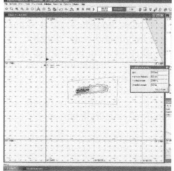

Figure 11. Oil slick thickness is 6.6 mm.

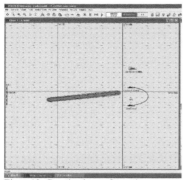

Figure12. Creating boom formation

Table 3: Response operation costs.

Report	time	date
begin	06:00	20.07.2010
end	15:43	20.07.2010

Costs by organizations

No organization

	Capitan Poinc	$7 112,60
	Expandi 4300 (800m)	$874,50
	m/s Czeslaw II	$2 176,53
	m/z Zefir	$796,77
	Ro-Boom 1500 (600m)	$1 739,28
	Seaskimmer 50	$871,50
Total		$13 571,18
		40713,55 PLN

5 CONCLUSION

A major use of this model could be for control of response resources (contingency planning) and find new locations for vessels and equipment to minimize costs of cleanup. It could be used for simulations and training. Further research needs to test the validity of such model.

Optimization of location response resources depending on reduction of costs is also very important. Full complement of planned simulations, based on predicted ships' accidents, should give an answer: whether an allocation of responses or their expansion are necessary. Protection of the Baltic Sea environment without bearing the unnecessary costs is a main purpose of research. First results of accident and antipollution action based on the real data are described in this paper.

ACKNOWLEDGEMENT

This paper was created with support of project Baltic Master II, partially EU founded Baltic Sea Region Programme 2007-2013.

REFERENCES

Galt J. A.: "The Integration of Trajectory Models and Analysis Into Spill Response Information Systems: the Need for a Standard" Second International Oil Spill Research and Development Forum, May 1995.

Gucma L., Przywarty M.: : "The Model of Oil Spills Due to Ships Collisions in Southern Baltic Area", Conference proceedings TRANS'NAV 2007 "Advances in Marine Navigation and safety of sea transportation", Monograph, Edited by Adam Weintrit, The Nautical Institute, Gdynia 2007, ISBN: 978-83-7421-018-8.

HELCOM "Overview of the ship traffic", April 2009

Lazuga K., Juszkiewicz W.:"The probability of cost pollution-accident simulation in the PISCES II Simulator", Conference proceeding 8th International Probabilistic Workshop, Szczecin 2010.

National Plan of Fighting Threats and Environmental Pollutions at Sea. SSAR, Gdynia 2005.

Specification for PISCES II, July 2007.

Oil Spill Response
Miscellaneous Problems in Maritime Navigation, Transport and Shipping – Marine Navigation and Safety of Sea Transportation – Weintrit & Neumann (ed.)

28. Modeling of Accidental Bunker Oil Spills as a Result of Ship's Bunker Tanks Rupture – a Case Study

P. Krata & J. Jachowski
Gdynia Maritime University, Poland

J. Montewka
Aalto University, Finland; Maritime University of Szczecin, Poland

ABSTRACT: An accidental bunker oil discharge, regardless of a ship type, is more common sort of accidents then widely considered cargo spills from tankers. However, there is a shortage of studies addressing this sort of incidents in terms of mathematical modeling, especially time-domain modeling. The case study presented in the paper comprises the three dimensional CFD-based (Computational Fluid Dynamics) modeling of bunker spills from ships due to hull and fuel tanks rupture. The time needed for bunker oil to be spilled and a rate of the spill are assessed as well as the total volume of spilled oil.

1 INTRODUCTION

The most commonly an oil accident is considered as a result of an oil tanker cargo spill. However this sort of disaster is rising the most the public awareness, hopefully cargo spills are not the most common ones (ITOPF, 2011). The most frequent oil spill accidents are those related to accidental discharges from bunker tanks of all vessels, not only tankers.

Contrary to tankers transporting heavy oil as cargo, ships' bunker tanks are not required to have a double hull structure as protection. The statistics on the oil spills provided by The International Tanker Owners Pollution Found reveal that the significant amount of oil spills of a size in a range 7-700 tones come from accidents like collisions and groundings (Fig. 1) (ITOPF, 2011).

In the past 40 years there were 1249 minor spill accidents, whereas 444 cases account for the major oil spills of a size above 700 tons, and only tankers are considered. Taking into account all types of ships this number can be expected significantly higher. These data shows that the minor size oil spills should not be neglected in any risk analysis related to maritime traffic, as they are the most frequent oil spills to occur.

At present there are two recognized and adopted methods for the bunker spill size estimation. Both of them seem to be quite general in nature and rough in results.

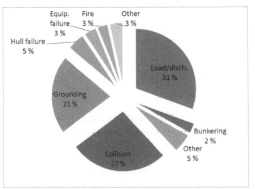

Figure 1. The distribution of causes of minor oil spills (7-700 tons) (ITOPF, 2011)

One approach is based on historical data. This approach is relatively easy, fast and straightforward, thus it has gained popularity among researches and was adopted in some general studies (HELCOM 1990, Safetec 1999, Gucma&Przywarty 2007, Nyman 2009). The limitations of this approach and its weak areas were pointed out by Michel&Winslow (1999) and Eide et al (2007), where the main concern about historical data was that they are not necessarily representative to today's accident scenarios, mostly due to changes in ship construction or layout of the tanks.

Another method, included in the IMO guidelines for approval of alternative tanker designs (IMO 1995, IMO 2003), contains a probabilistic-based procedure for estimating oil outflow performance. Probability density functions describing the location, extent and penetration of side and bottom damage are applied to a vessel's compartmentation, generating the probability of occurrence and collection of damaged compartments associated with each possible damage incident. All oil is assumed to outflow from tanks penetrated in collisions, whereas outflow from bottom damage is based on pressure balance calculations. This method sounds, however, more reliable than the previous it still lacks the time component. The method does not provide this vital information on the rate of the spill nor the time needed for tank to be released. From the preparedness and response point of view this parameter is essential, as the bunker spills occur in a close vicinity of a shore and the response time is usually very limited.

Recently a methodology has been introduced based on the analytical calculations and time domain simulations in order to calculate the volume of oil outflow and outflow rate versus time (Tavakoli et al. 2008, Tavakoli et al. 2010). The method addresses accidental cargo spills from tankers.

In this paper a method for bunker spill estimation in spatial-temporal domain is presented. The methodology takes into account the fluid dynamics, the size of a tank rupture is estimated with the use of the IMO methodology. However the damaged tank is assumed not to be a subject to longitudinal and transverse motions.

2 BUNKER OIL SPILL MODELING BY 3-D CFD METHODS

The technique for oil spill modeling applied in the paper makes use of Computational Fluid Dynamics. Authors propose the methodology aiming at estimation a quantity of the bunker spill, a rate of such a spill and time for the bunker to release. The method can contribute some information to the probabilistic approach utilized in previously mentioned IMO methodology. CFD based solution seems to be useful for better understanding the oil outflow process and its duration.

The proposed methodology has a wide range of applications and is free of the constraints typical for IMO statistical approach. In the paper a model for bunker spill estimation is put forward and finally a case study is presented, which is assumed as an exemplary grounding accident.

The 3-dimentional simulations of oil trickling and disseminating in water phenomenon were performed by the use of the commercial code "Fluent". The software is an universal and flexible tool designed for modeling of liquids dynamics. Most commercial CFD codes use the finite-volume or finite-element methods which are well suited for modeling flow past complex geometries (Bhaskaran&Collins). The Fluent code uses the finite-volume method (FVM), and uses the volume of fluid (VOF) method for free surface problems (Dongming&Pengzhi 2008, Fluent 2006).

The numerical simulations of the oil dispersing in water phenomenon were performed for a number of damage extend configuration (Fig.4) and tank geometry corresponding to the relevant parameters of the selected bulk carrier. The cross section of a vessel and the location of a damaged double bottom tank is shown in Figure 2.

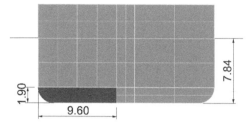

Figure 2. The cross section of a ship and her double bottom tank to be ruptured

In the course of the study a typical double bottom bunker tank of an exemplary bulk carrier is considered. The characteristic dimensions of the damaged tank are as follows:
- length – 40.0 m;
- breadth – 9.6 m;
- double bottom height – 1.9 m.

The shape of the double bottom bunker tank is presented in Figure 3.

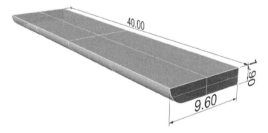

Figure 3. The damaged tank shape and dimensions

A leakage of bunker oil might take place due to a variety of reasons among which collision or grounding are the most common (Fig.1). There are available statistical analyses of damage locations in all three dimensions within ships hulls and damage extends. Usually collision and grounding are researched separately and nowadays some widely accepted distributions of hull damage size are in use.

Nevertheless the diversity of damage location is noticed, for the purpose of the case study one exemplary double bottom tank is considered and one elevation of the damage above the ship keel. The vertical extend of tank damage and its location is shown in Fig. 4.

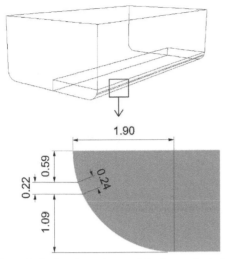

Figure 4. The location and span of the damage in double bottom tank (an exemplary case study)

The numerical simulations of oil spill were carried out for a number of tank damage lengths, i.e. 10%, 40%, 70%, 85% and 100% of compartment length (Fig. 5). This was to estimate how far nonlinear effects influence the final results, especially in terms of a rate of the outflow. The variable damage length was modeled by the use of removable panels concept which was convenient from the computational mesh creation point of view.

When the considered geometry of the damaged tank was established a set of assumptions required for the numerical simulations needed to be set up. The assumptions are related to the computational mesh creation, Courant number range, time step, fluid viscosity modeling, etc. Then the boundary conditions were defined.

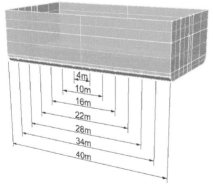

Figure 5. Removable panels used for modeling of different length shell damage

A variable computation time step was applied in the solution of the conservation equations for mass, momentum, and volume fraction of the both liquids water and oil. All numerical simulations were based on a 3D quadrilateral mesh created in GAMBIT. The setup of computational mesh is shown in Fig. 6.

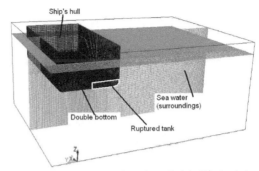

Figure 6. The computational mesh applied in 3D simulations performed by use of Fluent code

The computational mesh contains the considered section of the damaged ship (see Fig. 2 & 4) and a cuboid of surrounding water. The adjustment of mesh geometry is one of the key points of CFD modeling.

In the finite-volume method, such a quadrilateral is commonly referred to as a "cell" and a grid point as a "node". In this approach, the integral form of the conservation equations are applied to the control volume defined by a cell to get the discrete equations for the cell (Bhaskaran&Collins 2009).

The Fluent code is to find a solution such that mass, momentum, energy and other relevant quantities are being conserved for each cell. Also, the code directly solves for values of the flow variables at the cell centers; values at other locations are obtained by suitable interpolation (Bhaskaran&Collins 2009).

The Fluent code is designed to solve the Reynolds Averaged Navier Stokes (RANS) equations. RANS equations govern the mean velocity and pressure. These quantities vary smoothly in space and time, thus they can be relatively easy to solve; however they require some additional modeling to "close" the equations and these models introduce significant error into the calculation (Bhaskaran&Collins 2009).

In the course of the computation the variable time steps were applied in order to solve the conservation equations for mass, momentum, and volume fraction of the liquid. A concept of a VOF model is based on the monovalent assignment of liquid density inside every single computational cell (Fig. 7).

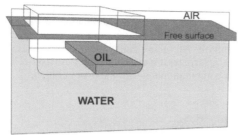

Figure 7. The mixture density concept in VOF model and mesh formulation for bottom damaged tank - at the time t=0s the oil starts to flow out of a crack (see Fig. 6 for clarification of the ruptured tank location)

Due to modeled fluids characteristics and relatively low value of expected Reynolds number, the laminar flow is applied in the course of numerical simulations. The laminar flows are characterized by pretty smoothly varying velocity fields in space and time in which individual layers move past one another without generating noticeable cross currents. These flows arise when the fluid viscosity is sufficiently large to damp out any perturbations to the flow (Bhaskaran&Collins 2009).

3 RESULTS OBTAINED IN THE COURSE OF CFD COMPUTATION - A CASE STUDY

The most straightforward attitude towards bunker oil spills consideration is just to carry out the series of simulations of an oil outflow. The methodology proposed in the paper is based on a CFD modeling. The simulations were performed on the basis of conditions and assumptions described in the previous sections of the paper.

The results of computations can be analyzed from variety viewpoints. The first outcome of the simulations is a visualization of flow patterns during the spilling process. This provides a general outlook on the considered phenomenon and helps to imagine the possible course of action in case of ship hull damage (Fig. 8).

The next result of CFD computation is a possibility of the velocity vector field visualization (Fig. 9). This is to facilitate the description of an oil outflow consecutive stages and its interpretation. In the Figure 9 the velocity vectors of the oil outflow from a tank can be noticed. Moreover, the velocity vectors of water flooding a tank may be also indentified.

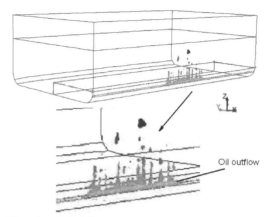

Figure 8. A typical flow pattern obtained for bunker oil outflow by the use of CFD simulations

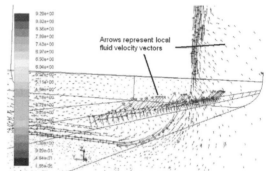

Figure 9. The velocity vector field visualization

However the main purpose of the CFD simulations was to estimate the volume of the oil spilled to the sea. It is important that the result obtained are time dependent. Thus the maximum allowed time for oil combating action may be assessed. The exemplary progress of the bunker oil outflow is shown in Figure 10. The percentage given in the graph refers to the length of hull damage according to the Figures 3 and 5.

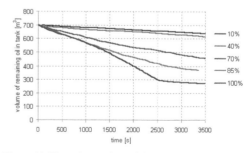

Figure 10. The volume of remaining oil in a damaged bunker tank (the initial volume of full tank was equal 700 m³)

The graphs plotted in Figure 10 seems to be rather smooth but one should keep in mind that they have a cumulative character describing the volume of oil remaining in the damaged tank. However, the rate of an oil spill is not a steady value while the oil trickling and disseminating in water phenomenon is not a stationary process.

The rate of oil discharge was computed and plotted for a time span of carried out CFD simulations. A number of graphs present the results of computations. The oil outflow rate obtained for the length of damage equals 25% of the section length is shown in Figure 11. And respectively: for 70% - in Figure 12, 85% - Figure 13 and 100% - Figure 14.

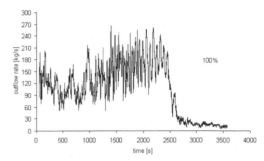

Figure 14. The rate of an oil outflow for the length of damage equal to 100% of considered section's length

The most typical feature of the computed oil discharge rate is its unstable character in terms of time. The hint to the explanation of this observation might be the flow pattern presented in Figure 8. The visualization reveals bubble-like character of the oil outflow resulting in variable value of the oil discharge rate.

All the results of performed CFD computations are rather coarse due to the adopted assumptions at the preliminary stage of the research. Thus, the upper limit of the Courant number was accepted relatively high and the computational mesh was generated not very dense. Such assumptions are justified for a feasibility study and obviously they will be modified for the planned main research purposes to obtain the expected satisfactory level of accuracy.

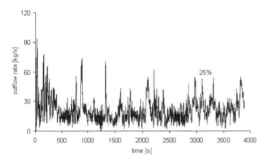

Figure 11. The rate of an oil outflow for the length of damage equal to 25% of considered section's length

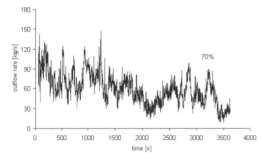

Figure 12. The rate of an oil outflow for the length of damage equal to 70% of considered section's length

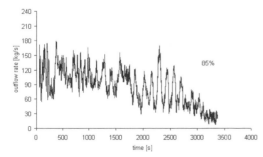

Figure 13. The rate of an oil outflow for the length of damage equal to 85% of considered section's length

4 SUMMARY

The study presented in the paper is a preliminary stage of the planned research and should be rather found as a practical approach to the feasibility study not the final result. However, a number of remarks and conclusions may be drawn.

First of all the realistic possibility of an application of CFD method to the bunker oil spill problem is revealed. The accuracy of computation may be improved by generation of larger size of meshes and lowering the limit of accepted Courant number.

From the point of view of shipping stakeholders the key point of the study is a remark, that CFD application enables estimation of bunker spill characteristics at the design stage of a ship. A variety of scenarios (different layouts of bunker tanks) can be examined and compared against the expected size of an oil spill.

Contrary to contemporary utilized methods, the method presented in this paper provides a number of advantages like time dependent characteristics of spilled oil volume and a rate of discharge. Such data might be useful also in the course of planning and conducting an oil combating action.

The flexibility of presented CFD-based approach benefits with strictly desirable proactive character of the method which is a good prospective for future research in the field of oil spills protection in the marine industry.

REFERENCES

Bhaskaran R., Collins L., Introduction to CFD Basics, Cornell University, New York, 2009, http://dragonfly.tam.cornell.edu/teaching/mae5230-cfd-intro-notes.pdf.

Dongming Liu, Pengzhi Lin, A numerical study of three-dimensional liquid sloshing in tanks, Journal of Computational Physics 227 (2008), pp. 3921–3939, 2008.

Eide, Magnus S., Øyvind Endresen, Øyvind Breivik, Odd Willy Brude, Ingrid H. Ellingsen, Kjell Røang, Jarle Hauge, and Per Olaf Brett, Prevention of oil spill from shipping by modelling of dynamic risk. Marine Pollution Bulletin 54, no. 10 (October), pp. 1619-1633, 2007.

FLUENT Tutorial Guide, Fluent Inc. 2006.

Gucma L., Przywarty M. 2007. Probabilistic method of ship navigational safety assessment on large sea areas with consideration of oil spills possibilities. In Taerwe&Proske (eds) 5th International Probabilistic Workshop, Ghent, pp.349-362, 2007.

HELCOM. 1990. Study of the Risk for Accidents and the Related Environmental Hazards from the Transportation of Chemicals by Tankers in the Baltic Sea Area. Baltic Sea Environment Proceedings No. 34. Helsinki Commission 1990.

IMO, Interim guidelines for approval of alternative methods of design and construction of oil tankers. Regulation 13F(5) of Annex I of MARPOL 73/78, Resolution MEPC.66(37), 1995.

IMO, Revised interim guidelines for the approval of alternative methods of design and construction of oil tankers. Regulation 13F (5) of Annex I of MARPOL 73/78 Resolution MEPC.110(49), 1993.

ITOPF, Information services / data and statistics. ITOPF. http://www.itopf.com/information-services/data-and-statistics/statistics/index.html#no, 2011.

Michel K., Winslow T., Cargo Ship Bunker Tanks: Designing to Mitigate Oil Spillage. SNAME Joint California Sections Meeting, 1999.

Nyman T. Evaluation of methods to estimate the consequence costs of an oil spill. 7 FP report, SKEMA study, 2009.

Safetec UK. Identification of Marine Environmental High Risk Areas (MEHRAs) in the UK. Doc. No.: ST-8639-MI-1-Rev 01. December 1999.

Tavakoli M.T., Amdahl J., Ashrafian A., Leira B.J. Analytical predictions of oil spill from grounded cargo tankers. Proceedings of the ASME 27th International Conference on Offshore Mechanics and Arctic Engineering, Estoril, Portugal, 2008.

Tavakoli M.T., Amdahl J., Leira B.J. Analytical and numerical modelling of oil spill from a side damaged tank. In Ehlers and Romanoff (eds) 5th International Conference on Collision and Grounding of Ships, Espoo 2010.

29. The Profile of Polish Oil Spill Fighting System

A. Bąk & K. Ludwiczak
Maritime University of Szczecin, Szczecin, Poland

ABSTRACT: Article presents the profile of polish oil spill fighting system along the polish coast. Its antipollution equipment and readiness for pollution fighting. Moreover the responsibilities and information flow are also presented. In the end authors give the reader conclusions after many of simulations performed in the Potential Incident Simulation, Control and Evaluation System (PISCES II) simulator which is installed in the Maritime University of Szczecin.

1 POLISH SAR BASES AND ITS SPILLAGE FIGHTING EQUIPMENT

Polish oil spill lighting system based on the same structure as SAR system. The main headquarter is located in Gdynia. The auxiliary one is situated in Świnoujście. Along the polish coast there are twelve regional SAR centres which are equipped with oil spill fighting means (Fig. 1). Depending on the oil spill area and its kind the proper centre or centres are designated to handle the situation.

Fig. 1. Deployment of polish SAR centres.
Source: www.sar.gov.pl

Spillage fighting equipment is distributed in SAR bases along the polish coast. At present (2010), the best equipped base is auxilliary base in Świnoujście, where the biggest polish rescue vessel m/v Kapitan Poinc is moored (Fig. 2). The readiness time for taking part in antispillage action is about 120 minutes. Moreover, there are other ships which are able to fight with oil pollution, ie.:
- SAR vessel type SAR – 1500 m/v Cyklon (Fig. 3),

- SAR vessel designated for oil spill fighting m/v Czesław II.

The equipment consists of many barriers and oil collectors, like:
- Barrier Expandi 4300 – 600 metres long,
- Barrier Seapack 80 – 450 metres long,
- Barrier Trellboom – 450 metres long,
- Oil collector Seaskimmer 50 – efficiency 50 m^3/h,
- Oil collector Walosep W2 – efficiency 45 m^3/h,
- Oil collector for heavy oil Scantrawl A.

In the rest of polish SAR bases the spillage fighting equipment is similar to that presented above. The main difference lay in another kind of ships and their worthiness against pollution. The detailed plan of its distribution is published in the Internet (http://www.sar.gov.pl).

Fig. 2. SAR vessel m/v „Kapitan Poinc". Photo: Andrzej Bąk.

As a basis of oil spill fighting action is to immediate countertact oil spillage, which in effect leads to water pollution. Every incident, no mater the origin (vessel, underwater pipeline, etc.), should be reported to Marine Pollution National Contact Point (MPNCP) located in SAR Centre Gdynia with no delay. Ports incidents should be reported to Harbour Master. Breakdowns of inland industrial installations, which can affect marine environment, have to be reported to Maritime Office.

Fig. 3. SAR vessel m/v „Cyklon". Photo: Andrzej Bąk.

MPNCP officer on duty establishes the following details:
- The nature of incident,
- Numbers of people onboard,
- Determine the threat for people and equipment,
- Type, dimensions and name of the vessel or any other installationtypu,
- Identification of owner or operator,
- Identification of position, course, speed, vessels in vicinity,
- Information about coastal installation, distances to shallow water and the coast,
- Cargo information, amount of bunker and indicate which of them are dangerous for marine environment,
- Constructional and mechanical integrity of vessel,
- Weather conditions and sea state,
- Required assistance,
- Making the attempt to avoid water pollution.

MPNCP officer on duty pass on the report to adequate Maritime Office. Its director makes a decision regarding incident in accordance with the rules. Whoever notices any water pollution or any threat to marine environment should inform MPNCP, by means of:
- Captain report, which was responsible for pollution,
- Captain report, which spotted the pollution,
- Airplane report, which was making patrol flight,
- Vessel pilot report,
- Report form any other airplane,
- Any person report, which spotted the pollution from the land,
- Coastal installation manger report,
- Harbour Master officer on duty report.

In case of small local oil spillage, adequate environmental protection inspector designated by Maritime Office director is taking command and checks if the action is performed according to regional antipollution plan. In the event of spillage which require to use national antipollution forces the director of Maritime Office is making decision of using SAR service. It is needed also to inform Ministry of Infrastructure and in danger of coast pollution the proper governor and provincial environmental protection inspector.

In the moment of beginning antipollution action the director of Maritime Office should inform the following entities:
- Ministry of Maritime Economy,
- Proper governor and provincial environmental protection inspector,
- Helsinki Commission Secretary,
- States the signatory of Helsinki Convention, if pollution can reach their coast,
- International Maritime Organization in case of very serious pollution.

Head of Operation cooperates with Polish Coastguard and SAR Coordination Centre. In order to have the overall view of the incident and to take proper decisions it is crucial to obtain actual data regarding pollution. The best way to do so is taking the observation and monitoring area affected by flying means. Thanks to that it is possible to search bigger area and indicate any other dangerous like additional spillage stains, possibility of laying down the oil substances ashore and threat of another country coast. Using airplanes have advantage of taking pictures and making photography documentation of polluted area. The person who is responsible for such a monitoring is Director of the Maritime Office in Gdynia who has at command TURBOLET airplane which can be supported by Navy planes [1].

2 SIMULATIONS AND CONCLUSIONS

In order to estimate the cost of antipollution action and optimal equipment distribution among all polish SAR bases authors made many simulations in the Potential Incident Simulation, Control and Evaluation System (PISCES II) simulator which is installed in the Maritime University of Szczecin. That simulator is often use for estimating the polluter and for monitoring the spillage [2]. As the result the following conclusions can be set:
- Proper choice of antipollution equipment speeds up the action itself,
- Heavy and light oil removing duration is comparable,
- Bed weather conditions like high sea, strong wind and current make the action very difficult and prolong the time of pollution fighting,

- Very often the bed weather conditions affects the laying down the oil substances ashore,
- In some cases at the Zatoka Gdańska it is necessary to use SAR vessel moored in Świnoujście as the best equipped in Poland nowaday,
- Duration of action can be many time long in bed weather which makes the cost of action very high,
- High sea (3 metres wave height or more) makes the action completely impossible, because the equipment is not designed for such a weather conditions,
- Cost of action can be reduced by equipping SAR bases with more efficient oil collectors and barriers. Also readiness time should be reduced as much as possible (at present it is 120 minutes).

BIBLIGRAPHY

[1] „Krajowy Plan Zwalczania Zagrożeń i Zanieczyszczeń Środowiska Morskiego", Morska Służba Poszukiwania i Ratownictwa, Gdynia 2005 (in Polish).
[2] Perkovic M., Sitkov A. „Oil Spill Modeling and Combat"

Large Cetaceans

30. Towards Safer Navigation of Hydrofoils: Avoiding Sudden Collisions with Cetaceans

H. Kato, H. Yamada, K. Shakata, A. Odagawa, R. Kagami & Y. Yonehara
Tokyo University of Marin Science and Technology (TUMSAT), Tokyo, Japan

M. Terada & K. Sakuma
KHI JPS Co., Ltd., Kobe, Japan

H. Mori & I. Tanaka
Kawasaki Heavy Industries Ltd., Kobe, Japan

H. Sugioka & M. Kyo
Japan Agency for Marine-Earth Science and Technology (JAMSTEC), Yokohama, Japan

ABSTRACT: Recently, sudden collisions between large cetaceans and high-speed hydrofoils have become problematic to Japanese sea transport in some localities. We therefore initiated a project to investigate approaches for minimizing risk to both ships and cetaceans. Under the present project, the following three subprojects are underway: clarifying which whale species are found near sea routes and determining their seasonal variations; identifying whale species that have a high collision risk; and determining the unique acoustic characteristics of high-collision-risk cetaceans for the improvement of underwater speakers (UWS). By conducting acoustic surveys using novel methods, including an anatomical approach based on characteristics of the inner ear, the aim of this project is to accurately estimate the audible range of species with a high collision risk and improve the sounds generated by the UWS. Thus far, we have identified the cetacean species at high-risk in two major sea routes. In the next phase of the study we plan to develop an imaging system that recognizes a cetacean's unique blow using an infrared camera, in an attempt to warn of the approach of high-collision-risk whale species at an early stage by sounding an alarm.

1 INTRODUCTION

Although the situation has improved somewhat recently, a series of collisions have occurred in some Japanese sea routes between hydrofoil-type super high-speed vessels (hereafter "HF vessels"; Fig. 1) and large marine life and continue to concern transport officials and other relevant peoples to this issue. These collisions are often the target of marine accident inquiries that examine the responsible factors, and it would be accurate to state that in most instances, the collisions involve cetaceans (Fig. 2). Thus, such collisions not only negatively impact safe navigation, but also represent a risk to the survival of cetaceans.

Figure 2. Sperm whale, a whale species with a high collision risk with hydrofoil-type super high-speed vessels. Their population in the waters around Japan has increased in recent years. (Photograph by H. Kato)

Beginning around the year 2000 in Europe, the Agreement on the Conservation of Cetaceans in the Black Sea Mediterranean Sea and Contiguous Atlantic Area (ACCOBAMS), which is based on the Bonn Convention on the Conservation of Migratory Species of Wild Animals (UNEP/CMS), was formed in response to increased concerns about collisions between large cetaceans and ships. In tandem, discussions were also initiated at the International Whaling Commission (IWC), which is the main organization for the management of cetacean stocks and whaling issues. In 2008, at the 60th IWC Annual Meeting held in Santiago, Chile, cooperation with the International Maritime Organization (IMO) was strength-

Figure 1. Hydrofoil type of high-speed vessel, which are used for important high-speed sea routes linking remote islands with the Japanese mainland. (Photograph by H. Kato)

ened, as proposed by the Netherlands, and this cooperation was also promoted at the 61st IWC Annual Meeting held in June 2009 in Madeira, Portugal. Under this cooperative effort, both international organizations organized the Joint IWC-ACCOBAMS Workshop on Reducing Risk of Collisions between Vessels and Cetaceans (Anon, 2010).

Against this background, this paper outlines our research project aimed at reducing risks to both ships and cetaceans against sudden collisions, particularly for large cetaceans, and discusses future directions in this field. The main sections of this paper have been taken from selected sections presented in a previous paper by Kato (2009).

2 WHY IS OUR RESEARCH PROJECT NECESSARY?

The IWC, ACCOBAMS, and IMO regard collisions between large cetaceans and super high-speed vessels as one of significant threats to the survival of cetaceans. Their main strategies for deterring such collisions are clear and simple, and involve requiring HF vessels to settle on the surface and reduce speed whenever whales appear, and establishing protection areas for whales. However, as these approaches will not always prevent collisions, additional research projects to identify more effective strategies are necessary.

Japan, which is completely surrounded by sea, has a population of approximately 130 million people. The Japanese population is predominantly concentrated in urban areas, whereas in rural and mountainous areas, extreme depopulation is occurring. The numerous islands surrounding Japan are no exception; the depopulation of the once well-developed islands will devastate them, leading to environmental damage and disturbances to the coastline, which will eventually penetrate the offshore areas. HF vessel services help limit the depopulation of the islands, as they drastically shorten the travel time between them and the mainland, making frequent travel more feasible. However, increases in the number and frequency of HF vessels on sea routes poses increased risk for collisions between vessels and cetaceans. . The main goal of our research project is to identify effective approaches for limiting the risks associated with collisions between large cetaceans and HF vessels.

3 OUTLINE AND PROGRESS OF THE RESEARCH PROJECT

In April 2006, the Maritime Bureau of the Ministry of Land, Infrastructure, Transport and Tourism established a committee for considering safety measures for HF vessels. The Laboratory for Ceta-

cean Biology at Tokyo University of Marine Science and Technology, based on the working group established by the above-mentioned committee, began conducting research on collision avoidance by seeking the cooperation of Kawasaki Shipbuilding Corporation, a maker of Jetfoils (JF) which are the main type of hydrofoil-type HF vessels, KHI JPS Co., Ltd., which is in charge of HF vessel maintenance, and several additional companies, such as Sado Kisen Co., Ltd., and Tokai Kisen Co., Ltd., which operate JF services.

A number of HF vessels are equipped with underwater speakers (UWS) that emit sound waves with the aim of repelling cetaceans. However, whales (86 species in total; 14 species within the suborder Mysticeti and 72 species within the suborder Odontoceti, as recognized by the Scientific Committee of the IWC in 2010) differ markedly in their acoustic characteristics depending on the species, particularly between baleen whales, which are highly adapted to the ocean, and toothed whales, which have retained numerous traits from their terrestrial mammal ancestors (Fig. 3).

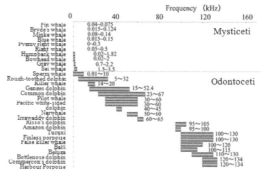

Figure. 3 Differences in the sonar frequency bands of cetacean species grouped at the sub-order level of taxon. Cited literature for sonar frequencies: Mysticeti (baleen whales), Au, 2000; Odontoceti (toothed whales), Backus and Schevill, 1966.

Due to the large variations in sonar frequency bands among cetaceans, questions remain concerning the efficacy of the currently used UWS. It is also not clear what type of sound each whale species can hear, particularly by large cetacean species, which are difficult to keep in captivity. However, working on the assumption that cetaceans are capable of hearing calls of their fellow whales, we propose identifying the hearing range of high-collision-risk cetacean species using a novel method.

To this end, the following two sub-projects have been established:

1 Clarify which whale species are found near shipping routes and determine their seasonal variations to identify whale species that have a high collision risk, and then reflect their unique acous-

tic characteristics in UWS. To achieve this goal, we have initiated the following sub-projects.
- Analysis of visual data typically collected by ship crews in service
- Visual surveys to determine whale species by whale specialists
- Improvement in the accuracy of typical visual data collection by visual training of ship crews to identify whale species
- Improvement in the identification accuracy of cetacean species and detection ability through the introduction of high-definition video cameras. Identify whale species with a high collision risk based on surveys of whale body size and the above research results, and then modify the sounds generated by UWS for each route and season
2 Conduct acoustic surveys using novel approaches, with the aim of estimating the audible range of whale species with a high collision risk.
- Estimations using an anatomical approach of the inner ear
- Estimations from correlations with vocal characteristics
- Further improvement of the sounds generated by UWS based on the above findings

With regards to sub-project (1), significant progress has been made, and the results have been compiled in two Master's theses submitted to Graduate School, Tokyo University of Marine Science and Technology (Odagawa, 2007; Shakata, 2008).

In addition to the analyses presented in these two theses, we collected further survey data until 2010. Through analyses of visual data typically collected and data from specialized cetacean sightings on the major HF vessel sea routes from Tokyo to Okada on Izu-Ohsima Island and from Niigata to Ryotsu on Sado Island (Fig. 4), we were able to identify the high-risk species and their peak season for predicting the most probable months for collisions (Table 1).

Table 1. High-risk cetacean species involved in collisions with HF vessels and their peak season associated with selected sea routes.

Sea route	Critical species	Season
Tokyo – Okada (Izu-Ohshima Island)	Sperm whale (*Physeter Macrocephalus*)	Year round with peak in Oct.-Dec
	Baird's beaked whale (*Berardius bairdii*)	May-Aug.
Niigata – Ryotsu (Sado Island)	Minke whale (*Balaenoptera acutorostrata*)	Year round with peak in Apr.-May

Table 2. Estimated vocal frequencies for sperm and Bryde's whales based on vocalizations collected from field surveys using a hydrophone (from Yamada et al., 2011).

Survey location	Species	Vocal frequency (kHz)
Tosa Bay	Bryde's whale (*Balaenoptera edeni*), Mysticeti (baleen whales)	0.15-0.40
Bonin Island	Sperm whale (*Physeter macrocephalus*) Odontoceti (toothed whales)	1.90-4.80

With respect to sub-project (2), we have been conducting field surveys for the collection of natural vocalizations of several large cetaceans. The results of this work are described in the report by Yamada *et al.* (2011), submitted to the same volume as the separate dedicated paper.

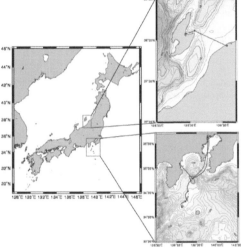

Figure 4. Map showing the two major HF vessel sea routes for Tokyo - Izu Ohshima Island (bottom right) and Niigata - Sado Island (top right).

Due to logistical limitations to access large cetaceans, we prioritized two field sites to collect vocalizations from Bryde's whales (*Balaenoptera edeni*) and sperm whales, which were located in the neritic waters off Kochi (Tosa Bay; approx. 33N–133E) and in the waters off the southern coast of Chichijima Island (Bonin Islands; 27N-142.2E), respectively, which are promising locations to access these cetacean species. As reported in Yamada *et al.* (2011), the observed ranges of vocal frequencies were 1.90-4.80 kHz and 0.15-0.40 kHz for sperm and Bryde's whales, respectively (Table 2). As both upper values for the vocal frequencies of the two species are higher than the reported values presented in Fig. 1, this finding may allow the signals emitted by UWS to be modified for increased efficacy.

Although vocal frequencies provide indirect evidence of the audible (frequency) range, more direct measurements clearly represent a more suitable approach in the context of UWS. Yamada *et al.* (2011) also reported the predicted audible range for several large cetaceans determined using an anatomical approach of the inner ear: 0.12-15.00 kHz and 0.11-31.10 kHz for the common minke and Baird's beaked whale, respectively (Table 3). Due to differences in the species examined, the values estimated using this approach cannot be directly compared to those of the vocal frequency estimated in the field surveys. However, as similarities in vocal characteristics are generally expected among related taxon groups (Fig. 1), comparisons among Balaenoptera species (common minke whale - Bryde's whales) would appear to be valid.

Table 3. Predicted audible ranges for the common minke and Baird's beaked whales based on anatomical approaches using the inner ear (from Yamada et al., 2011).

Sampling location	Species	Predicted frequency (kHz)
Off Ishinomaki, Miyagi, Japan (Pacific coast)	Common minke whale *(Balaenoptera acutorostrata)* Mysticeti (baleen whales)	0.12 - 15.93
Off Wadaura Chiba, Japan (Pacific coast)	Baird's beaked whale *(Berardius bairdii),* Odontoceti (toothed whales)	0.27 - 33.09

From the reported literature, it was confirmed that the audible range is wider than that of the vocal frequency range, indicating that cetaceans can hear sounds of higher frequency than of their own vocalizations, at least among Balaenoptera cetaceans and also likely among Mysticeti cetaceans. Although the obtained evidence is not strong enough to reach a firm conclusion, this nature of cetaceans (i.e, audible range > vocal frequency range) is likely also true among Odontoceti species, as determined from the comparison between the vocal frequency of sperm whales and the audible range of Baird's beaked whale. As a further step for reducing risk of collisions between HF and cetaceans, this rationale can be implied for the improvement of UWS.

4 SHORT-TERM PROPOSAL FOR IMPROVE-MENT OF UWS

To minimize risks for both HF vessels and cetaceans, we are presently preparing interim proposals for improving specifications of UWS by taking into account the results of the present project; these improvements are ongoing and are halfway complete.

First, it is necessary to adjust the sounds generated by UWS using appropriate acoustic specifications of the high-risk cetacean species for which sudden collisions are to be avoided. We have already identified the critical cetacean species to be sperm whales and Bride's whales in the Tokyo-Izu-Ohshima sea route, and common minke whales in the Niigata-Ryotsu sea route. For the seas routes in other localities, identification of the critical species through cetacean sighting surveys is an essential first step.

For the improvement of UWS, it is of critical importance to identify the most effective sound frequency to repel the critical cetaceans. In this regard, Yamada *et al.* (2011) suggested that existing UWS (operating at 6-20 kHz) should be modified to produce frequencies less than 15 kHz for common minke whales, less than 0.4 kHz for Bryde's whales, and less than 4.8 kHz for middle to large-toothed whales, such as sperm whales. However, it is unknown whether such modification is technically feasible, or if installation of the necessary hardware is realistic. Therefore, more investigations are needed to better estimate the audible range for critical cetacean species, and to determine the required technical and mechanical improvements of UWS. Particularly, an increased number of anatomical inner ear samples is expected to help clarify the audible range and effective frequency for deterring high-collision-risk cetacean species.

5 FUTURE CHALLENGES AND POTENTIAL APPLICATIONS

In the course of our investigations and research for the improvement of UWS, specific challenges for future research have been identified. We previously attempted to introduce high-definition cameras to improve the accuracy of identifying whale species, but a new technique for obtaining high-quality images has emerged that might contribute to the early detection of whales, as well as allow the more accurate identification of cetacean species.

In a potential application utilizing this imaging technique, we are considering constructing a system that recognizes a whale's unique blow in an image to warn of the approach of high-collision-risk cetacean species at an early stage by sounding an alarm by UWS. We have already initiated the implementation of such a system, as outlined in a pilot study by Yonehara *et al.* (2011). Although some collisions are inevitable, considering the user friendliness of current HF vessels and the high ability of crew members to steer and maneuver ships, we believe that if an approaching cetacean is detected in advance, the frequency of collisions with large cetaceans can be drastically reduced.

The positions of the HF vessel manufacturers and operating companies regarding this issue have been very positive, and further efforts are being made in the pursuit of greater safety for both passengers and cetaceans.

ACKNOWLEDGEMENTS

Our sincerely thanks are due to cooperation with the managers, officers, and ship crew of the HF vessel operating companies, Sado Kisen Co., Ltd. and Tokai Kisen Co., Ltd. We also thank the many officers of the Maritime Bureau of the Ministry of Land, Infrastructure, Transport and Tourism for their initial guidance.

REFERENCES

Anonymous. 2010. Report of the joint IWC-ACCOBAMS workshop on reducing risk of collisions between vessels and cetaceans. *Report of the joint IWC-ACCOBAMS workshop on ship strikes* :1-42.

Au, W.W.L. 2000. Hearing in whales and dolphins; An overview. In W. W. L. Au, A. N. Popper & R. R. Fay *(eds.) Hearing by whales and dolphins*: 1-42. New York: Springer.

Backus, R. and W.E. Schevill. 1966. "Physeter clicks," in Whales, Porpoises and Dolphins. *In* K. S. Norris *(ed.) Whales, dolphins, and porpoises.* : 510-528. California: Berkeley.

Kato, H. 2009. Towards Avoiding Collisions between Super High-speed Vessels and Whales. *The Ship & Ocean Newsletter.*217:4 — 5.

Odagawa, A. 2007. Fundamental research of collision avoidance between cetaceans and super high-speed vessels. *Master Thesis, Tokyo University of Marine Science and Technology* :1-105.

Shakata, K. 2009. Fundamental research of collision avoidance between cetaceans and super high speed vessels - Analysis for the occurrence of cetaceans on the sea routes and increase in precision of cetacean identification by the introduction of video camera research-. *Master thesis, Tokyo University of Marine Science and Technology* :1-90.

Yamada, H., L. Kagami, Y. Yonehara, H. Matsunaga, M. Terada, K. Okanoya, T. Kawamoto & H. Kato. 2011. Estimated audible range of large cetacean for improvement of the Under Water Speaker. *The TransNav2011 monographs.*

Yonehara, Y., L. Kagami, H. Yamada, H. Kato, M. Terada & S. Okada. 2011. Feasibility study for infrared detecting of large cetaceans to avoid sudden collision. *The TransNav2011 monographs.*

31. Estimation on Audibility of Large Cetaceans for Improvement of the Under Water Speaker

H. Yamada, L. Kagami, Y. Yonehara, H. Matsunaga & H. Kato
Tokyo University of Marine Science and Technology, Tokyo, Japan

M. Terada
KHI JPS Co., Ltd., Kobe, Japan

R. Takahashi
Kawasaki Heavy Industries Ltd., Kobe, Japan

K. Okanoya
RIKEN Brain Science Institute, Wako, Japan

T. Kawamoto
Tsurumi University, Yokohama, Japan

ABSTRACT: In order to avoid collisions between the hydrofoil (HF) and cetaceans, the Under Water Speaker (UWS) has been installed on the HF. Because of its potential in utility, we tried to improve the UWS to minimize the risk of the collisions. Under our project, we examined three subprojects; 1) Analyzing the characteristics of the HF underwater noise; 2) Assessing audibility of major large cetaceans by measuring their vocalizations and 3) An anatomical prediction of the audible range by examining the cochlear basal membrane. Through the analyses, it was identified that the noise produced by the HF was a broad-band noise with approximately 150dB *re* 1μPa-m. That noise level was lower than those of larger boats suggesting difficulties for cetaceans in sensing approach of the vessels. In addition, analysis of their vocalizations and anatomical obervation indicated that dominant frequency of their audible range was lower than signals produced by the existing UWS.

1 INTRODUCTION

The Under Water Speaker (UWS) has been installed on the hydrofoils (HF) for avoiding the collisions with large cetaceans. However, its utility is still uncertain whether the sound produced by the UWS corresponds to the audible range of major large cetaceans. This is a major reason why we conduct the present study which explores the way to improve the UWS from biological aspects. Under the present research project, we examined three sub-projects:

1.1 Characteristics of the HF underwater noise

One of the reasons of the collision is considered that the HF underwater noise is possibly hard for cetaceans to recognize approaching vessel. It is probably because the noise level is too low and hardly transmits to a long distance. Therefore we analyze the characteristics of the HF underwater noise.

1.2 Assessing audibility by measuring of vocalization

The UWS should be improved to prevent the collision incorporating with the audible range of causal cetaceans. Currently, there are no direct measures of audible range for any large cetaceans because they

cannot be investigated with conventional audiometric techniques of psychoacoustical or electrophysiological analysis. However, the audible range can be assessed by vocalization, as to correspond the dominant frequencies of the vocalization (e.g. calls) to the most sensitive region of receptor system in vertebrate taxa (Green and Marler 1979). Shakata *et al.* (2008) identified sperm whale (*Physeter macrocephalus*), Baird's beaked whale (*Berardius bairdii*), common minke whale (*Balaenoptera acutorostrata*), Bryde's whale (*Balaenoptera edeni*) as possible causal species of the collision on the sea route of the HF in Japanese water. Among these species, we chose to sample the vocalization of sperm whale and Bryde's whale since relatively easier to record their vocalizations in Japanese water. Based on the recorded vocalization, we assessed the audible range of these species.

1.3 Anatomical Predictions of the Audible range

Alternatively, a comparative anatomy approach is the useful way to estimate the audible range because anatomical structure of inner ear correlates to frequency range in multiple mammalian species (Echteler *et al.*, 1994). In particular, the cochlear configuration and thickness to width (T/W) ratios of the basilar membrane in inner ear are consistent with the maximal and minimum frequencies for each cetacean species (Ketten and Wartzok, 1990). This study estimates the audible range of common minke whales and Baird's beaked whale by describing the anatomy of their inner ears and applying the model described by Ketten (2000).

2 METHODS

2.1 *Characteristics of the HF underwater noise*

Underwater noise of the HF, SUISEI: 169gt. LOA31.2m (Owned by Sado Kisen Co.,Ltd.), was recorded during its cruise at service speed (38-39kn) from a small vessel at a distance of 100m. Recordings were made using a OKI SEATEC model OST2130 (frequency response 10Hz to 100kHz) omnidirectional hydrophone has sensitivity of approximately -174±3dB *re* 1V/μPa with 10m cable. It was connected via pre-amplifiers (frequency response from 20Hz to 20kHz), on a Sony PCM-D50 digital recorder (16bit 44.1 kHz) and OKI SEATEC OST4100 Hydroacoustic analyzer which was used to analyze the sound source level. This recording chain had a flat frequency response from 20Hz to 20 kHz. The HF underwater noise was assessed by 1/3-octave bands analysis using Avisoft SASLab Pro (Avisoft Bioacoustics, Germany.Ver.4.1.) because noise levels in 1/3-octave bands are useful in interpreting noise effects on animals. The estimated source levels of underwater noise (at 1m) of the HF were calibrated by Transmission Loss and Absorption Loss (Francois & Garrison1982).

2.2 *Assessing audibility by measuring of vocalization*

Bryde's whale sounds were recorded in the waters of Kochi on the south western coast of Japan (32°40' to 33°2'N, 133°00' to 133°13'E) for five days in mid-October, 2008. The study area ranged from the south coasts out to approximately 30km (16 nmi) of the shore. We chartered a fishing-boat for recording. When cetaceans were sighted, the boat approached to confirm species and school size and to collect other relevant information. When sighting Bryde's whale, the hydrophone was thrown in water and started recording. Signals were recorded with a OKI SEATEC model OST2130 omnidirectional hydrophone with 15m cable, connected via pre-amplifiers (frequency response 20Hz to 20kHz), on a Sony PCM-D50 digital recorder(16bit 44.1 kHz). This recording chain had a flat frequency response from 20Hz to 20 kHz. The acoustic characteristics of phrases were examined by using the analysis software Avisoft SASLab Pro, with spectrogram parameters of 512-point FFT size, 75.0% overlap, and Hamming window. The vocalization was analyzed based on the following parameters; duration, peak frequency, and fundamental frequency of element.

Sperm whale sounds were recorded off the southeastern coast of Chichijima, the Bonin (Ogasawara) Islands (26°55' to 27 °05 'N, 142°11' to 142°24'E) for eight days in September, 2009. We chartered a fishing-boat for recording. When sperm whales were sighted the boat approached to confirm school size and to collect other relevant information. When sighting sperm whale, the hydrophone was thrown in water and their vocalization was recorded. Signals were recorded with recording system described above in Bryde's whale sounds recording.

2.3 *Anatomical Predictions of the audible range*

Ear bones of 9 specimens of common minke whales (9 individuals) and 6 of Baird's beaked whales (3 individuals) were collected (under cooperation with The Institute of Cetacean Research, Tokyo Japan and National Research Institute of FarSeas Fisheries, Yokohama Japan) and analyzed. Ears were frozen shortly after the collection and placed in a buffered 10% formalin solution. All ears were scanned by the nuclear magnetic resonator (NMR) (Bruker Bio Spin AVANCE 400WB) to measure the cochlear configuration. The ears were decalcified in 5% formic acid for three weeks and processed into slides 10-μm cryosections by the Kawamoto film-sectioning method (Cryofilm transfer kit; *Leica* Microsystems) (Kawamoto 2003). Every 10th section was stained with hematoxylin and eosin and mounted. Basilar membranes were shown by a laser scanning microscope Olympus Model FV1000 at a ×10 (width) × 20 (thickness) objective magnifications with a scale and ocular calibrated scale for measurements. The basilar membranes were measured for width and thickness using ImageJ (National Institutes of Health, USA. Ver.1.43.).

3 RESULTS

3.1 *Characteristics of the HF underwater noise*

Underwater noise of the HF was a "broadband" sound with energy spread continuously over a range of frequencies.

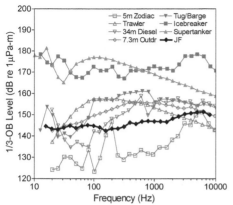

Figure 1. Estimated 1/3-octave source levels of underwater noise (at 1m) of the HF and other vessels summaries of Richardson *et al.*(1995).

Source levels at 1m were estimated by cylindrical spreading transmission loss TLc = 10log r (dB) and absorption loss (Francois & Garrison1982) with distance from source (100m), water depth (88.8m), water temperature (19°C), salinity (35‰), pH (8). As a result, the estimated source level was 146.3±2.6 dB *re* 1µPa-m(Mean±SD) with peak sound level of 151.4 dB *re* 1µPa-m at 6,300Hz. The sound level of the HF was almost equal to that of small ships (Fig.1).

3.2 *Assessing audibility by measuring of vocalization*

48 biological sounds estimated to be emitted by Bryde's whale were recorded during a total of 8h24m15s recording time. We judged whether sounds were emitted by Bryde's whale based on the following two points, 1) any marine animals other than Bryde's whale were not visually-observed during recordings, 2) these sounds showed similarities to Bryde's whale vocalizations described by Oleson et al. (2003). These sounds were assigned to two categories: a) swept tonal call, b) harmonics call (Fig.2).

1 Swept tonal call [Fig.2(a)] was detected 46/48 calls. Table 1 indicates a summary of the quantitative parameters of this call type. These calls were tonal and frequency modulated sounds characterized by an arch-like structure and no repetition. The mean peak frequency of these calls was 269.9Hz±71.3 (mean±SD) ranging from 131.6Hz to 373.4Hz. The mean duration for this call type was 0.71 s±0.30 (mean±SD). This type calls were first recorded off the coast of Japan.

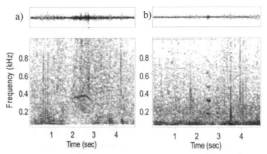

Figure 2. Envelope curves and spectrograms of two phrase types attributed to Bryde's whales in Japan. (a) Swept tonal call (b) Harmonics call. Both spectrograms were made with a 512-point FFT, 75.0% overlap, and Hamming window.

2 Harmonics call [Fig.2 (b)] was detected only 2/48 calls in this study. The calls included higher-frequency harmonics [fundamental frequency 78.5(74.0-83.0) Hz] than these reported by Oleson *et al.* (2003) (approximately 45Hz). The mean of duration for this call type was 0.28 (0.17-0.39) s.

A total of 12547 clicks of sperm whales were recorded during a total of 7h20m23s recording time (Fig.3). Table 1 indicates a summary of the quantitative parameters of clicks. The peak frequency of the clicks was 3174Hz (geometric mean, 95% CI 3140-3208). The duration of the individual pulses within a click is 9.27±0.05ms(Mean±SD). The recorded levels of the clicks were approximately 150 dB *re* 1µPa.

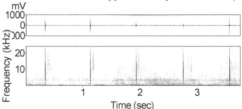

Figure 3. Envelope curves and spectrograms of the clicks of sperm whales. Spectrograms were made with a 512-point FFT, 75.0% overlap, and Hamming window.

Table1. Frequency quantitative parameters for vocalization.

Species (Whale)	Sound type	Frequency Range(Hz)	Peak Frequency(Hz)
Bryde's whales	Swept tonal call (n=46)	131.6-373.4	269.9±71.3 (Mean±SD)
	Harmonics call (n=2)	250.0-293.0	271.5 (Mean)
Sperm Whales	Clicks (n=12581)	1870-4780 (3140-3208)	3174
			(GM, 95% CI)

3.3 *Anatomical Predictions of the audible range*

Initial surveys of cochlear dimensions from NMR images showed that common mike whales cochlear were type M while Baird's beaked whales cochlear were type II (Ketten 2000) (Fig.4a). Furthermore we measured the cochlea length and other cochlea configurations shown in Table 2. It took approximately 3 weeks to complete decalcification of the cochlear. The Kawamoto film-sectioning method allowed the best preparation of thin sections from specimens of the cochlear (Fig 4b). All specimens had measurable intact basilar membranes in apex and base region of the cochlea(Fig 4c). Table 2 shows the thickness/width ratios and estimated frequency of the audible range for each species from the data using the model described in Ketten (2000).

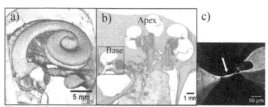

Figure 4. Images of cochlea from Baird's beaked whales. a) A three-dimensional reconstruction by NMR. b) Images from histology slide preparations. c) The basilar membrane(arrow) of the cochlea basal turn (20×)

Table 2. The cochlear spiral and the basilar membrane measurements, and predicted frequency of the audible range from the measurements [the model described in Ketten (2000)].

Species	Common minke whales (n=9, 9individuals)	Baird's beaked whales (n=6, 3individuals)
Number of turns	2.32(±0.09)	2.08(±0.09)
Membrane Length(mm)	54.82(±2.20)	54.44(±2.35)
Basal diameter (mm)	12.36(±0.83)	16.14(±2.35)
Axial height (mm)	7.36(±0.55)	7.66(±1.09)
Membrane Thickness Base/Apex (μm)	9.0/5.4	15.9/13.5
Membrane Width Base/Apex (μm)	171.4/1128.0	142.4/304.5
T/W ratio Base/Apex	0.0525/0.0098	0.1568/0.020
Predicted Frequency (kHz)	15.93/0.12	33.09/0.27

4 DISCCUSION

Large cetaceans response to sound level higher than from 110 to 170dB *re* 1μPa (Richardson *et al* 1995), and it requires 170dB re 1μPa to trigger a strong reaction when they are away from the source (Akamatsu 1993). Since the HF cruising sound level at 100m from source had 126.3dB *re* 1μPa (source level 146.3 (±2.6) dB *re* 1μPa-m) was probably too low to make whales react to the sound. In addition, peak frequency of the HF may be higher (6.3 kHz) than sensitive hearing of large cetaceans. Therefore, it is necessary to install the UWS that effectively produces sounds that make whales recognize the approaching the HF.

Because it is to correspond the dominant frequencies of the vocalization to the most sensitive region of receptor system in vertebrate taxa (Green and Marler 1979), the present study assessed the dominant audible ranges for each whale as follows; Bryde's whales 0.1-0.4kHz, sperm whales1.9-4.8kHz.

Alternatively, an anatomical structure of inner ear correlates to the maximal and minimum frequency of the audible range in each cetacean species (Ketten and Wartzok, 1990). Therefore the audible ranges for each whale were predicted as follows; common minke wahles: 0.1-15.9kHz and Baird's beaked whales:0.06-33.1kHz.

Thus, it is considered that the existing the UWS (6-20kHz) is necessary to be modified to produce the lower frequency down to less than 15.9kHz for common minke whales, to less than 0.4kHz Bryde's whales, and between 1.9 to 4.8kHz for sperm whale.

As for Baird's beaked whales, predicted audible rangea are well inside of those by the existing UWS. However the vocal frequency for Bryde's whales is fur below of to lower band by the UWS. The gaps are thought to be technically difficult to fill up. Because the frequency of vocalization is certainly within the audible range and the practical audible range is much wider, this must be investigated by further examination through anatomical approach mention above.

For further study, it is necessary to improve the acoustic property of the UWS based on the sound known to have a repellent effect against large cetaceans within the frequency range shown in this study.

5 CONCLUSION

The HF noise level was probably too low to make whales react. Therefore, it is necessary to install the UWS effectively. Based on vocalizations and anatomical observation, it is considered that the existing the UWS (6-20kHz) is necessary to be modified to produce the lower frequency down to less than 15.9kHz for common minke whales, to less than 0.4kHz Bryde's whales, and between 1.9 to 4.8kHz for sperm whale.

ACKNOWLEDGMENT

We would like to special thank for supporting for investigation Sado Kisen Co., Ltd. Ougata-cyou Yugyo-sensyukai (Whale watching association), Ogasawara whale watching association and Ogasawara Marine Center. Special thanks to Ms. M. Yamaguchi and Mr. A. Nakajima, who helped us recording of sperm whales. Key specimens provided by the Institute of Cetacean Research and National Research Institute of FarSeas Fisheries. Thanks to Laboratory of Maine Biology, Tokyo University of Marine Science and Technology and Biolinguistics,RIKEN Brain Science Institute. This work supported by Kawasaki Heavy Industries Co.,Ltd and KHI JPS Co., Ltd.,.

REFERANCES

Akamatsu, T., Hatakeyama, Y. & Takatsu, N. 1993. Effects of pulse sounds on escape behavior of false killer whales. *Nippon Suisan Gakkaishi.* 59(8):1297-1303.

Echteler, S.W., Fay, R.R & Popper, A.N. 1994. Structure of the mammalian cochlea. In: A.N. Popper.(eds), *Comparative hearing: mammals* :134-171.New York. Springer-Verlag. Press.

Francois, R.E. & Garrison, G. R. 1982. Sound absorption based on ocean measurements: Part I: Pure water and magnesium sulfate contributions. *Journal of the Acoustical Society of America.* 72(3): 896-907.

Green, S & Marler, P. 1979. The analysis of animal communication. In: P. Marler & J.G. Vandenbergh. (eds), *Social behavior and communication, vol 3. Handbook of behavioral neurobiology*: 73-158. New York. Plenum Press.

Kawamoto, T. 2003. Use of new adhesive film for the preparation of multi-purpose fresh-frozen sections from hard tissues, whole-animals, insects and plants. *Arch. Histol. Cytol* .66(2):123-143.

Ketten, D.R & Wartzok, D. 1990. Three-dimensional reconstructions of the dolphin ear. In: R. Kastelein.(eds), *Sensory abilities of cetaceans.* :81-105.New York: Plenum Press.

Ketten, D.R. 2000. Cetacean ears. In: R.R. Fay. (eds), *Hearing by whales and dolphins*: 43–108.New York. Springer-Verlag Press.

Oleson, E.M., Barlow, J., Gordon, J., Rankin, S., & Hildebrand, J. A. 2003. Low frequency calls of Bryde's whales. *Marine Mammal Sci.* 19: 160–172.

Richardson, W.J. 1995. Marine mammal hearing. In:W. Richardson,C.R.Greene, C.I.Malme & D.H. Thomson.(eds), *Marine Mammals and Noise.*:205-240. California. Academic press.

Shakata, K., Odagawa, A., Yamada, H., Matsunaga, H., & Kato, H. 2008. Toward to avoiding ship strike of cetaceans with the high-speed HF(1)-Identifying expected cetacean species on the track lines of HF. *2008 Annual meeting of M.S.J. summary* :135

32. Feasibility on Infrared Detection of Cetaceans for Avoiding Collision with Hydrofoil

Y. Yonehara, L. Kagami, H. Yamada & H. Kato
Tokyo University of Marine Science and Technology, Tokyo, Japan

M. Terada & S. Okada
KHI JPS Co., Ltd., Kobe, Japan

ABSTRACT: To achieve safer navigation without sudden collisions with large cetaceans at high speed boats such as the hydrofoil, we examined its feasibility of an installation of the infrared camera. Because any cetaceans are of air-breathing animals, it is theoretically expected that they can be potentially detected through imaging of the infrared cameras. Thus, we examined the feasibility of detection with aiming at sperm whales in waters off Chichijima Islands (27°4'N, 142°13'E), Japan. Through the experiment, it was revealed that sperm whales could be detected stably within 200m, and detectable cue were blow, back body and fluke tails. However, boats and waves were also detected as noise images. Especially, waves greatly resemble the whale back bodies. Although potential of the infrared camera was confirmed, there are still necessities of further experiments including ones conducting at different temperate waters, to successfully install the infrared camera for earlier finding of large cetaceans.

1 INTRODUCTION

Concerning the problem of collision accident between large cetaceans and vessels, it is expected to identify the critical season and species of large cetacean in some courses of hydrofoils off Japan water and also to technically improve the Under Water Speaker (UWS) for repellent of cetaceans. However, it is necessary to take measures of not only repellent of cetacean but also actively cetacean avoidance. The UWS is one of the countermeasures for cetaceans to give way. For the safety navigation of hydrofoils, as part of these studies, we are envisaging an alert system triggered by the whale detection using infrared camera (thermography). Because any cetaceans are of air-breathing animals, they come to the surface inevitably in regular interval and their blow or body parts appear at the sea surface. In this system, we aim at these objects as detection cue by using infrared camera.

In this study, we examine the feasibility of the infrared camera on detecting large cetaceans. On that basis, we intended to form the foundation of early cetacean detection and alert system using infrared camera by formulating the necessary conditions or the method of infrared camera operation for detecting whales.

2 MATERIAL AND METHOD

2.1 *Equipments*

In this experiments, we used a Infrared Thermal camera; Thermo Tracer TH9260 made by NEC/Avio Infrared Technologies (Fig. 1). Its detector is uncooled focal plane array (microbolometer), and thermal image pixels is 640(H) × 480(V). The operating band is 8 to 13 μm, and it can measure objects from -40°C to 500°C. Optical field of view of this camera is fixed (21.7°(H)×16.4°(V)) . This camera obtains the images 30frame per second.

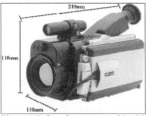

Figure 1 Infrared camera used in this study (TH9260)

Additionally, we used HD video camera (Sony Digital HD video camera recorder HDR-SR1) for recording normal images and thermal images in the same time as a control of the experiment. For measuring accurate distance between cetaceans and us, we used the Laser range finder (Bushnell Laser Rangefinder Elite 1500).

2.2 Experiment of infrared detection on sperm whales in Bonin Water

An infrared recording experiment of sperm whales was conducted off the South-East water off Chichijima island, Ogasawara Islands (also known as the Bonin Islands, a subtropical archipelago located ca. 1000 km directly south of central Tokyo in the western Pacific Ocean) from September 12 to 14, 2009 (Fig. 2).

We conducted this survey by chartered small fishing boat "Shoeimaru" (Fig. 3). At eye level was approximately 2m since TH9260 was held at the bow. Captain and three researchers got on the boat and participate in this survey. We intended to investigate sperm whales mainly and record boats or sea surface at the same time. We left Futami port every morning and returned before sunset. The investigation time was less than 10 hours a day.

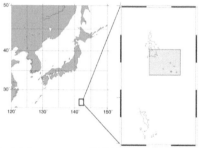

Figure 2 A survey area off Chichijima island (Bonin Islands). Gray square indicates survey area (26°55′ to 27°05′ N, 142°11′ to 142° 24′ E). Cross marks indicates the position sperm whale sighted.

Figure 3 Syoeimaru; Length:12m,

After leaving port, we began a search of sperm whales by eyes. When we found the sperm whale, we tried to approach them and began the IR recording experiment in diverse distance. A researcher held the IR camera. Others served the distance measurement and data notation, or filming the behavior of sperm whale and the research status by the HD camera. When the blow or body of sperm whales appeared above sea surface, we noted whether researcher could perceive it on the display of IR camera (Success or Failure) with visual assistance of sperm whale behavior made by other researchers and an experienced operator of "Shoeimaru" in sperm whale sightings around the waters. And we also noted the time, type of the object and distance from the object at the same time. We saved a thermal image or thermal movie if at all possible. The distance from the object was measured by use of Laser range finder. If it couldn't, researcher measured the distance with the eye.

Acquired infrared data were conducted an analysis of temperature measurement using dedicated software (NS9205Viewewr program made by NEC/Avio infrared technology). We measured temperatures of objects as T_a and temperatures of sea surface around the objects as T_{sea} (Fig. 4) and calculated the difference in temperature ($\triangle T = T_a - T_{sea}$).

Figure 4 The example of measuring method that of T_a and T_{sea} by using software (This is the magnified figure). In this figure, point 'a' indicates the temperature of T_a and 'b' indicates that of T_{sea}.

3 RESULTS

We could conduct the experiment in three days. Table1 shows the summary of the experiments. In this experiment, we could observe and obtain data from Sperm whale, small boat and sea surface. Show the result of observation below.

Number of IR detection indicates the number of data that we noted whether or not to perceive sperm whales on IR camera in diverse distances. We excluded the data that noted same result (S or F), same distance and within three minutes to avoid data bias. However, it has potentially observed same individual, other cases of data were dealt with the different results since it is very hard to identify that the whale examined previously or not. Number of thermal images indicates the whole number of saved thermal images through the experiment.

Table 1 Summary of the experiment in waters off Ogasawara, 2010.

Date	12 Sept.	13 Sept.	14 Sept.
Observed whale groups	2	2	2
Number of IR detection	21	8	33
Number of thermal images	184	794	1371

3.1 Sperm whale

In this experiment, sperm whale's blow, back body and tail flukes were detected by using TH9260 (Fig. 5). As for object range, we could always detect these sperm whale objects on the display of TH9260 within a range of 150m. However, we never detect them out of a range of 350m. Within a range of 160m to 300m, we could detect or not (Table 2 and Fig. 5). The maximum distance of laser range finder was 118m.

Figure 5 Thermal images of sperm whale's objects detected by TH9260. Objects were indicated by arrows.
Top: blow (100m away), Middle: dorsal part of body (117m away) and Bottom: tail fluke (200m away).

Table 2 The result of IR sperm whale detection in each distance.

Distance (m)	Success	Failure	Sum
~50	6	0	6
51~100	12	0	12
101~150	12	0	12
151~200	11	6	17
201~250	2	7	9
251~300	3	1	4
301~	0	2	2
	46	16	62

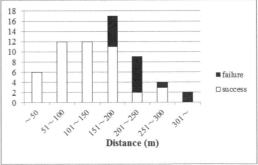

Figure 6 Numbers of success or failure in IR detection experiment each distances.

Concerning measured temperature, Median T_a and $\varDelta T$ of sperm whale obtained by TH9260 was 29.2°C and 1.30°C (Table 3 and Fig. 7). Median $\varDelta T$ of each sperm whale's detected objects were 1.10°C (blow), 1.40°C (back body), and 1.30°C (tail fluke) (Table 4, Fig.8). As a result, $\varDelta T$ of sperm whale's body parts indicates high value compared with that of their blow, however statistical significant difference didn't shown in Tukey-Kramer's multiple comparisons (P=0.185) between $\varDelta T$ of sperm whale objects.

Table 3 T_a and $\varDelta T$ estimated in each objects.

	T_a of each objects (°C)		
	Whale (n=41)	Boat (n=8)	Wave (n=13)
Median	29.2	33.9	29.4
Max	32.2	38.8	30.3
Min	25.8	32.7	27.1
	$\varDelta T$ of each objects (°C)		
	Whale (n=41)	Boat (n=8)	Wave (n=13)
Median	1.3	6.3	1.2
Max	2	11.3	1.5
Min	-0.3	5.1	0.4

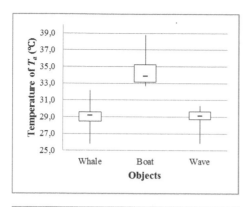

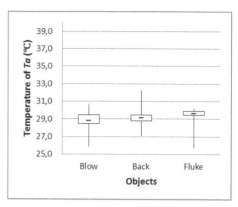

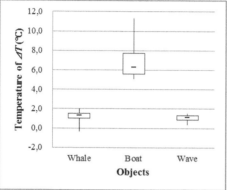

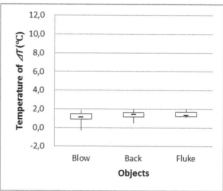

Figure 7 Box-plot of T_a and $\varDelta T$ of each objects
Vertical bar indicates the maximum and minimum value. Box indicates lower and upper quartile point. Short crossbar indicates median value.

Figure 8 Box-plot of T_a and $\varDelta T$ of sperm whale's objects
Vertical bar indicates the maximum and minimum value. Box indicates lower and upper quartile point. Short crossbar indicates median value.

Table 4 T_a and $\varDelta T$ estimated in each objects

T_a of sperm whale's objects (°C)			
	Blow (n=19)	Back (n=16)	Fluke (n=6)
Median	28.8	29.15	29.65
Max	30.7	32.2	30.2
Min	25.9	27.1	25.8
$\varDelta T$ of sperm whale's objects (°C)			
	Blow (n=19)	Back (n=16)	Fluke (n=6)
Median	1.1	1.4	1.3
Max	1.9	2	2
Min	-0.3	0.5	1.1

3.2 Small boat

We could observe a small boat employed for whale watching. The boat appeared very strong on the display of TH9260 since it has high rate of infrared radiation (Fig. 9). Median T_a and $\varDelta T$ of small boat was 33.4°C and 6.30°C (Table 3 and Fig. 7). It was highest temperature in this study, and statistical significant difference was showed in Tukey-Kramer's multiple comparisons (P<0.01) between $\varDelta T$ of boat and sperm whale or wave.

Figure 9 A small boat detected by TH9260.

3.3 *Sea surface*

In the display of TH9260, concentric temperature gradient always appeared on the sea surface (e.g. Fig. 7 and 11). Waves such as white caps were also detected on the display of TH9260 (e.g. Fig. 7). As for temperature, median T_a and $\varDelta T$ of waves was 29.10°C and 1.10°C (Table3 and Fig. 7).

4 DISCUSSION

4.1 *Detectable range*

In this experiment, it is suggested that maximum surely detectable range of TH9260 was 150m and stable detectable range of it seems to be approximately 200m (Table 2 and Fig. 6). However, it is considered that the measured instances aren't enough to conclude it. In early studies, McCafferty *et al.* (2007) introduced that Infrared camera has the potential for measuring mammals as far as 1000m or more. Barber *et al.* (1991) could detect walruses on the sea surface away from more than 2000m above in the sky. Thus, infrared camera might be able to detect large cetaceans at more long distance. The possible way to improve the detectable range of infrared cameras is using telephotographic lens and higher resolution infrared camera.

4.2 *Ta and $\varDelta T$ estimated in each objects*

Cuyler *et al.* (1992) made a survey of large cetaceans and taking their temperature using infrared camera in the northern Norway water (68~80°N). They resulted that $\varDelta T$ of sperm whale's blow and tail fluke was +3.00°C and +6.00°C. The tendency that body parts exceed blow in $\varDelta T$ corresponds to Cuyler *et al.* (1992). However, the $\varDelta T$ of our survey showed lower temperature than that of Cuyler *et al.* (1992). It is suggested that this caused by the difference of the regions that Cuyler *et al.* (1992) investigated in high latitude water (air temperature

was 2.5 ~13.0°C and water temperature was 2.7~10.1°C during investigation period), in contrast, we investigated in subtropical water (air temperature was 28.1~33.5°C and water temperature was 27.0~30.5°C during investigation period). Thus, it is considered that surrounding temperature make a large effect on cetacean detection using infrared camera. If this idea is correct, it is suggested the cetacean detection using infrared camera will be more effective in cool winter season at the sea off Japan. Especially, early detection of large cetaceans and alert system using infrared camera is expected to be more effective in the winter Sea of Japan which is hard to find cetaceans by the naked eye.

In our experiment, we also could detect a small boat, sea surface and waves through the thermal images as a noise of cetacean detection. The boat is thought to be able to distinguish from cetacean objects easily on the thermal images since it was indicated that the boat has large difference of temperature between sea surface and it emitted definitely high thermal energy than other objects (Table 3, and Fig. 5 and 9). Concerning concentric temperature gradient on sea surface and the temperature indicated by waves, it is inconsiderable that they reflect the real variation in water temperature. Though Infrared has directionality, it is suggested that apparent variation in temperature was occurred by changing the shooting angle between infrared camera and objects of shooting. Additionally, the configuration and the T_a composition of waves were similar to these of the sperm whale's back body in thermal images (Figs. 5, 7 and 8, and Table 3, 4). Therefore, it seems to be difficult to distinguish sperm whale's back body from waves in thermal images. It is suggested that cetacean detection using infrared camera will be hard to operate in heavy weather which waves appear a lot on the sea surface.

5 CONCLUSION

From this study, it was revealed that the infrared camera has the feasibility to detect large cetaceans on the sea around Japan. However, it is difficult to say that we obtained a sure result for the detectable range since the shortage of data. Therefore, it is necessary to accumulate more data of detectable range.

It is suggested that oceanic condition and air or water temperature affect the result of cetacean detection using infrared camera. Thus, we should conduct more elaborate survey of environmental factor that infrared cetacean detection works effectively. It is also concluded that sperm whales are difficult to distinguish from waves in thermal images. Therefore, we should devise the technique to distinguish them not only by temperature.

In addition, early study, Cuyler *et al.* (1992) conducted cool water and obtained more pronounced difference in temperature between cetaceans and sea surface than our study that conducted subtropical water. Thus, summer Bonin island water might be too warm to investigate the feasibility of infrared camera to detect large cetaceans, and it is necessary to make more practical verification surveys in an environment similar to the course of hydrofoils off Japan.

ACKNOWLEDGEMENT

This study was supported by KHI JPS Co., Ltd. and Kawasaki Shipbuilding Corporation. We wish to thank M. Yamaguchi and A. Nakajima for getting on the boat and carrying out the experiment with us. We are grateful to Ogasawara Marine Center and Ogasawara Whale Watching Association for support conducting the survey off Chichijima, Ogasawara (Bonin Islands).

REFERENCES

Barber, D. G., Richard, P. R., Hochheim, K. P., Orr J. 1991. Calibration of aerial thermal infrared imagery for walrus population assessment. *Arctic*, 44, 58-65.

Cuyler, L. C., Wiolsrod, R., Oritsland, N. A. 1992. Thermal infrared radiation from free living whales. *Marine Mammal Science*, 8(2): 120-134.

McCafferty, D. J. 2007. The value of infrared thermography for research on mammals: previous applications and future direction. *Mammal Rev.* 2007, Volume 37, No. 3, 207-233.

Shakata, K., Odagawa, A., Yamada, H., Matsunaga, H., Kato, H. 2008. Toward to avoiding ship strike of cetaceans with the high-speed HF (1) Identifying expected cetacean species on the track lines of HF. 2008 Annual meeting of M. S. J. summary: 135

Author index